Ideas and Opinions

by

Albert Einstein

Based on MEIN WELTBILD,
edited by Carl Seelig,
and other sources

New translations and revisions
by Sonja Bargmann

A CONDOR BOOK
SOUVENIR PRESS (EDUCATIONAL & ACADEMIC) LTD

Ideas and Opinions title verso

First published in the USA by Bonanza Books, New York

Copyright © 1954 by Crown Publishers Inc.

First published in Great Britain, 1973, by Souvenir Press Ltd
43 Great Russell Street, London WC1B 3PD

Reissued 2005

Reprinted 2005

ISBN 0285647253

Printed and bound in Great Britain by MPG Books, Bodmin, Cornwall

CONTENTS

PUBLISHER'S NOTE

Ideas and Opinions represents an attempt to gather together, so far as is possible, in one volume the most important of Albert Einstein's general writings. Until now there have been three major collections of articles, speeches, statements, and letters by Einstein: *The World As I See It,* translated by Alan Harris, published in 1934; *Out of My Later Years* (1950), containing material from 1934 to 1950; and *Mein Weltbild,* edited by Carl Seelig, published in Switzerland in 1953, which contains certain new materials not included in either of the other collections.

Ideas and Opinions contains in the publisher's opinion the most important items from the three above-mentioned books, a few selections from other publications, and new articles that have never been published in book form before. It was only with the very kind cooperation of Carl Seelig and Europa Verlag of Zurich and the help of Professor Einstein himself that it was possible to assemble this collection of Einstein's writings from the earliest days to addresses of only a few weeks ago.

Special thanks must be given to Helen Dukas who facilitated the gathering of these articles and to Sonja Bargmann whose contribution is major: she checked and revised previous translations, provided new translations for all other articles not specifically credited, participated in the selection and editing of the entire volume and influenced her husband, Valentine Bargmann, to write the introduction to Part V, *Contributions to Science.*

Acknowledgments should be made also to the various publishers who made available articles copyrighted by them and whose names may be found with the articles.

PART I

IDEAS AND OPINIONS

PARADISE LOST

Written shortly after the establishment in 1919 of the League of Nations and first published in French. Also published in Mein Weltbild, *Amsterdam: Querido Verlag, 1934.*

As late as the seventeenth century the savants and artists of all Europe were so closely united by the bond of a common ideal that cooperation between them was scarcely affected by political events. This unity was further strengthened by the general use of the Latin language.

Today we look back at this state of affairs as at a lost paradise. The passions of nationalism have destroyed this community of the intellect, and the Latin language which once united the whole world is dead. The men of learning have become representatives of the most extreme national traditions and lost their sense of an intellectual commonwealth.

Nowadays we are faced with the dismaying fact that the politicians, the practical men of affairs, have become the exponents of international ideas. It is they who have created the League of Nations.

MY FIRST IMPRESSIONS OF THE U. S. A.

An interview for Nieuwe Rotterdamsche Courant, *1921. Appeared in* Berliner Tageblatt, *July 7, 1921.*

I must redeem my promise to say something about my impressions of this country. That is not altogether easy for me. For it is not easy to take up the attitude of impartial observer when one is received with such kindness and undeserved respect as

I have been in America. First of all let me say something on this score.

The cult of individuals is always, in my view, unjustified. To be sure, nature distributes her gifts unevenly among her children. But there are plenty of the well-endowed, thank God, and I am firmly convinced that most of them live quiet, unobtrusive lives. It strikes me as unfair, and even in bad taste, to select a few of them for boundless admiration, attributing superhuman powers of mind and character to them. This has been my fate, and the contrast between the popular estimate of my powers and achievements and the reality is simply grotesque. The awareness of this strange state of affairs would be unbearable but for one pleasing consolation: it is a welcome symptom in an age which is commonly denounced as materialistic, that it makes heroes of men whose goals lie wholly in the intellectual and moral sphere. This proves that knowledge and justice are ranked above wealth and power by a large section of the human race. My experience teaches me that this idealistic outlook is particularly prevalent in America, which is decried as a singularly materialistic country. After this digression I come to my proper theme, in the hope that no more weight will be attached to my modest remarks than they deserve.

What first strikes the visitor with amazement is the superiority of this country in matters of technology and organization. Objects of everyday use are more solid than in Europe, houses much more practically designed. Everything is designed to save human labor. Labor is expensive, because the country is sparsely inhabited in comparison with its natural resources. The high price of labor was the stimulus which evoked the marvelous development of technical devices and methods of work. The opposite extreme is illustrated by over-populated China or India, where the low price of labor has stood in the way of the development of machinery. Europe is halfway between the two. Once the machine is sufficiently highly developed it becomes cheaper in the end than the cheapest labor. Let the Fascists in Europe, who desire on narrow-minded political grounds to see their own particular countries more densely populated, take heed of this. However, the anxious care with which the United States keep out foreign goods by means of prohibitive tariffs certainly contrasts oddly with the general picture. . . . But an innocent visitor must not be expected to rack his brains too much, and when all is said and done, it is not absolutely certain that every question admits of a rational answer.

The second thing that strikes a visitor is the joyous, positive attitude to life. The smile on the faces of the people in photographs is symbolical of one of the greatest assets of the American. He is friendly, self-confident, optimistic—and without envy. The European finds intercourse with Americans easy and agreeable.

Compared with the American the European is more critical, more self-conscious, less kind-hearted and helpful, more isolated, more fastidious in his amusements and his reading, generally more or less of a pessimist.

Great importance attaches to the material comforts of life, and equanimity, unconcern, security are all sacrificed to them. The American lives even more for his goals, for the future, than the European. Life for him is always becoming, never being. In this respect he is even further removed from the Russian and the Asiatic than the European is.

But there is one respect in which he resembles the Asiatic more than the European does: he is less of an individualist than the European—that is, from the psychological, not the economic, point of view.

More emphasis is laid on the "we" than the "I." As a natural corollary of this, custom and convention are extremely strong, and there is much more uniformity both in outlook on life and in moral and esthetic ideas among Americans than among Europeans. This fact is chiefly responsible for America's economic superiority over Europe. Cooperation and the division of labor develop more easily and with less friction than in Europe, whether in the factory or the university or in private charity. This social sense may be partly due to the English tradition.

In apparent contradiction to this stands the fact that the activities of the State are relatively restricted as compared with those in Europe. The European is surprised to find the telegraph, the telephone, the railways, and the schools predominantly in private hands. The more social attitude of the individual, which I mentioned just now, makes this possible here. Another consequence of this attitude is that the extremely unequal distribution of property leads to no intolerable hardships. The social conscience of the well-to-do is much more highly developed than in Europe. He considers himself obliged as a matter of course to place a large portion of his wealth, and often of his own energies, too, at the disposal of the community; public opinion, that all-powerful force, imperiously demands it of him. Hence the most important cultural functions can be left to private enterprise and the part played by the government in this country is, comparatively, a very restricted one.

The prestige of government has undoubtedly been lowered considerably by the Prohibition law. For nothing is more destructive of respect for the government and the law of the land than passing laws which cannot be enforced. It is an open secret that the dangerous increase of crime in this country is closely connected with this.

There is also another way in which Prohibition, in my opinion, undermines the authority of the government. The public house is a place which gives people the opportunity to exchange views and ideas on public affairs. As far as I can see, such an opportunity is lacking in this country, the result being that the Press, which is mostly controlled by vested interests, has an excessive influence on public opinion.

The overestimation of money is still greater in this country than in Europe, but appears to me to be on the decrease. It is at last beginning to be realized that great wealth is not necessary for a happy and satisfactory life.

In regard to artistic matters, I have been genuinely impressed by the good taste displayed in the modern buildings and in articles of common use; on the other hand, the visual arts and music have little place in the life of the nation as compared with Europe.

I have a warm admiration for the achievements of American

institutes of scientific research. We are unjust in attempting to ascribe the increasing superiority of American research work exclusively to superior wealth; devotion, patience, a spirit of comradeship, and a talent for cooperation play an important part in its successes.

One more observation, to finish. The United States is the most powerful among the technically advanced countries in the world today. Its influence on the shaping of international relations is absolutely incalculable. But America is a large country and its people have so far not shown much interest in great international problems, among which the problem of disarmament occupies first place today. This must be changed, if only in America's own interest. The last war has shown that there are no longer any barriers between the continents and that the destinies of all countries are closely interwoven. The people of this country must realize that they have a great responsibility in the sphere of international politics. The part of passive spectator is unworthy of this country and is bound in the end to lead to disaster all round.

REPLY TO THE WOMEN OF AMERICA

Einstein's response to the protest of a women's organization against his visit to the United States. Published in Mein Weltbild, *Amsterdam: Querido Verlag, 1934.*

Never yet have I experienced from the fair sex such energetic rejection of all advances; or if I have, never from so many at once.

But are they not quite right, these watchful citizenesses? Why should one open one's doors to a person who devours hard-boiled capitalists with as much appetite and gusto as the Cretan Minotaur in days gone by devoured luscious Greek maidens, and on top of that is low-down enough to reject every sort of war, except the unavoidable war with one's own wife? Therefore give heed to your clever and patriotic womenfolk and remem-

ber that the Capitol of mighty Rome was once saved by the
cackling of its faithful geese.

THE WORLD AS I SEE IT

Originally published in Forum and Century, *Vol. 84, pp.
193–194, the thirteenth in the* Forum *series, "Living Philosophies." Included also in* Living Philosophies *(pp. 3–7),
New York: Simon and Schuster, 1931.*

How strange is the lot of us mortals! Each of us is here for
a brief sojourn; for what purpose he knows not, though he
sometimes thinks he senses it. But without deeper reflection
one knows from daily life that one exists for other people—
first of all for those upon whose smiles and well-being our own
happiness is wholly dependent, and then for the many, unknown
to us, to whose destinies we are bound by the ties of sympathy.
A hundred times every day I remind myself that my inner and
outer life are based on the labors of other men, living and dead,
and that I must exert myself in order to give in the same measure as I have received and am still receiving. I am strongly
drawn to a frugal life and am often oppressively aware that I
am engrossing an undue amount of the labor of my fellow-men.
I regard class distinctions as unjustified and, in the last resort,
based on force. I also believe that a simple and unassuming life
is good for everybody, physically and mentally.
    I do not at all believe in human freedom in the philosophical
sense. Everybody acts not only under external compulsion but
also in accordance with inner necessity. Schopenhauer's saying,
"A man can do what he wants, but not want what he wants,"
has been a very real inspiration to me since my youth; it has
been a continual consolation in the face of life's hardships, my
own and others', and an unfailing well-spring of tolerance.
This realization mercifully mitigates the easily paralyzing sense
of responsibility and prevents us from taking ourselves and

other people all too seriously; it is conducive to a view of life which, in particular, gives humor its due.

To inquire after the meaning or object of one's own existence or that of all creatures has always seemed to me absurd from an objective point of view. And yet everybody has certain ideals which determine the direction of his endeavors and his judgments. In this sense I have never looked upon ease and happiness as ends in themselves—this ethical basis I call the ideal of a pigsty. The ideals which have lighted my way, and time after time have given me new courage to face life cheerfully, have been Kindness, Beauty, and Truth. Without the sense of kinship with men of like mind, without the occupation with the objective world, the eternally unattainable in the field of art and scientific endeavors, life would have seemed to me empty. The trite objects of human efforts—possessions, outward success, luxury—have always seemed to me contemptible.

My passionate sense of social justice and social responsibility has always contrasted oddly with my pronounced lack of need for direct contact with other human beings and human communities. I am truly a "lone traveler" and have never belonged to my country, my home, my friends, or even my immediate family, with my whole heart; in the face of all these ties, I have never lost a sense of distance and a need for solitude—feelings which increase with the years. One becomes sharply aware, but without regret, of the limits of mutual understanding and consonance with other people. No doubt, such a person loses some of his innocence and unconcern; on the other hand, he is largely independent of the opinions, habits, and judgments of his fellows and avoids the temptation to build his inner equilibrium upon such insecure foundations.

My political ideal is democracy. Let every man be respected as an individual and no man idolized. It is an irony of fate that I myself have been the recipient of excessive admiration and reverence from my fellow-beings, through no fault, and no merit, of my own. The cause of this may well be the desire, unattainable for many, to understand the few ideas to which I have with my feeble powers attained through ceaseless struggle. I am quite aware that it is necessary for the achievement of the objective of an organization that one man should do the thinking and directing and generally bear the responsibility. But the led must not be coerced, they must be able to choose their leader. An autocratic system of coercion, in my opinion, soon degenerates. For force always attracts men of low morality, and I believe it to be an invariable rule that tyrants of genius are succeeded by scoundrels. For this reason I have always been passionately opposed to systems such as we see in Italy and Russia today. The thing that has brought discredit upon the form of democracy as it exists in Europe today is not to be laid to the door of the democratic principle as such, but to the lack of stability of governments and to the impersonal character of the electoral system. I believe that in this respect the United States of America have found the right way. They have a President who is elected for a sufficiently long period and has sufficient powers really to exercise his responsibility. What I value, on the other hand, in the German political system is the more extensive provision that it makes for the individual in case of illness or need. The really valuable thing in the pageant of human life seems to me not the political state, but the creative, sentient individual, the personality; it alone creates the noble and the sublime, while the herd as such remains dull in thought and dull in feeling.

This topic brings me to that worst outcrop of herd life, the military system, which I abhor. That a man can take pleasure in marching in fours to the strains of a band is enough to make me despise him. He has only been given his big brain by mistake; unprotected spinal marrow was all he needed. This plague-spot of civilization ought to be abolished with all possible speed. Heroism on command, senseless violence, and all the loathsome nonsense that goes by the name of patriotism—how passionately I hate them! How vile and despicable seems war to me! I would rather be hacked in pieces than take part in such an abominable business. My opinion of the human race is high enough that I believe this bogey would have disappeared long ago, had the sound sense of the peoples not been systematically corrupted by commercial and political interests acting through the schools and the Press.

The most beautiful experience we can have is the mysterious. It is the fundamental emotion which stands at the cradle of true art and true science. Whoever does not know it and can no longer wonder, no longer marvel, is as good as dead, and his eyes are dimmed. It was the experience of mystery—even if mixed with fear—that engendered religion. A knowledge of the existence of something we cannot penetrate, our perceptions of the profoundest reason and the most radiant beauty, which only in their most primitive forms are accessible to our minds— it is this knowledge and this emotion that constitute true religiosity; in this sense, and in this alone, I am a deeply religious man. I cannot conceive of a God who rewards and punishes his creatures, or has a will of the kind that we experience in ourselves. Neither can I nor would I want to conceive of an individual that survives his physical death; let feeble souls, from fear or absurd egoism, cherish such thoughts. I am satisfied with the mystery of the eternity of life and with the awareness and a glimpse of the marvelous structure of the existing world, together with the devoted striving to comprehend a portion, be it ever so tiny, of the Reason that manifests itself in nature.

THE MEANING OF LIFE

Mein Weltbild, *Amsterdam: Querido Verlag, 1934.*

What is the meaning of human life, or, for that matter, of the life of any creature? To know an answer to this question means to be religious. You ask: Does it make any sense, then, to pose this question? I answer: The man who regards his own life and that of his fellow creatures as meaningless is not merely unhappy but hardly fit for life.

THE TRUE VALUE OF A HUMAN BEING

Mein Weltbild, *Amsterdam: Querido Verlag, 1934.*

The true value of a human being is determined primarily by the measure and the sense in which he has attained liberation from the self.

GOOD AND EVIL

Mein Weltbild, *Amsterdam: Querido Verlag, 1934.*

It is right in principle that those should be the best loved who have contributed most of the elevation of the human race and human life. But if one goes on to ask who they are, one finds oneself in no inconsiderable difficulties. In the case of political, and even of religious, leaders it is often very doubtful whether they have done more good or harm. Hence I most seriously believe that one does people the best service by giving them some elevating work to do and thus indirectly elevating them. This applies most of all to the great artist, but also in a lesser degree to the scientist. To be sure, it is not the fruits of scientific research that elevate a man and enrich his nature, but the urge to understand, the intellectual work, creative or receptive. Thus, it would surely be inappropriate to judge the value of the Talmud by its intellectual fruits.

ON WEALTH

Mein Weltbild, *Amsterdam: Querido Verlag, 1934.*

I am absolutely convinced that no wealth in the world can help humanity forward, even in the hands of the most devoted worker in this cause. The example of great and pure individuals

is the only thing that can lead us to noble thoughts and deeds.
Money only appeals to selfishness and irresistibly invites abuse.

Can anyone imagine Moses, Jesus, or Gandhi armed with the
money-bags of Carnegie?

SOCIETY AND PERSONALITY

Mein Weltbild, *Amsterdam: Querido Verlag, 1934.*

When we survey our lives and endeavors, we soon observe
that almost the whole of our actions and desires is bound up
with the existence of other human beings. We notice that our
whole nature resembles that of the social animals. We eat
food that others have produced, wear clothes that others have
made, live in houses that others have built. The greater part of
our knowledge and beliefs has been communicated to us by
other people through the medium of a language which others
have created. Without language our mental capacities would
be poor indeed, comparable to those of the higher animals; we
have, therefore, to admit that we owe our principal advantage
over the beasts to the fact of living in human society. The individual, if left alone from birth, would remain primitive and
beastlike in his thoughts and feelings to a degree that we can
hardly conceive. The individual is what he is and has the significance that he has not so much in virtue of his individuality,
but rather as a member of a great human community, which
directs his material and spiritual existence from the cradle to
the grave.

A man's value to the community depends primarily on how far
his feelings, thoughts, and actions are directed toward promoting the good of his fellows. We call him good or bad according to his attitude in this respect. It looks at first sight as if our
estimate of a man depended entirely on his social qualities.

And yet such an attitude would be wrong. It can easily be
seen that all the valuable achievements, material, spiritual, and
moral, which we receive from society have been brought about

in the course of countless generations by creative individuals. Someone once discovered the use of fire, someone the cultivation of edible plants, and someone the steam engine.

Only the individual can think, and thereby create new values for society, nay, even set up new moral standards to which the life of the community conforms. Without creative personalities able to think and judge independently, the upward development of society is as unthinkable as the development of the individual personality without the nourishing soil of the community.

The health of society thus depends quite as much on the independence of the individuals composing it as on their close social cohesion. It has rightly been said that the very basis of Graeco-European-American culture, and in particular of its brilliant flowering in the Italian Renaissance, which put an end to the stagnation of medieval Europe, has been the liberation and comparative isolation of the individual.

Let us now consider the times in which we live. How does society fare, how the individual? The population of the civilized countries is extremely dense as compared with former times; Europe today contains about three times as many people as it did a hundred years ago. But the number of leading personalities has decreased out of all proportion. Only a few people are known to the masses as individuals, through their creative achievements. Organization has to some extent taken the place of leading personalities, particularly in the technical sphere, but also to a very perceptible extent in the scientific.

The lack of outstanding figures is particularly striking in the domain of art. Painting and music have definitely degenerated and largely lost their popular appeal. In politics not only are leaders lacking, but the independence of spirit and the sense of justice of the citizen have to a great extent declined. The democratic, parliamentarian regime, which is based on such independence, has in many places been shaken; dictatorships have sprung up and are tolerated, because men's sense of the dignity and the rights of the individual is no longer strong enough. In two weeks the sheeplike masses of any country can be worked up by the newspapers into such a state of excited fury that men are prepared to put on uniforms and kill and be killed, for the sake of the sordid ends of a few interested parties. Compulsory military service seems to me the most disgraceful symptom of that deficiency in personal dignity from which civilized mankind is suffering today. No wonder there is no lack of prophets who prophesy the early eclipse of our civilization. I am not one of these pessimists; I believe that better times are coming. Let me briefly state my reasons for such confidence.

In my opinion, the present manifestations of decadence are explained by the fact that economic and technologic developments have highly intensified the struggle for existence, greatly to the detriment of the free development of the individual. But the development of technology means that less and less work is needed from the individual for the satisfaction of the community's needs. A planned division of labor is becoming more and more of a crying necessity, and this division will lead to the material security of the individual. This security and the spare time and energy which the individual will have at his disposal can be turned to the development of his personality. In this way the community may regain its health, and we will hope that future historians will explain the morbid symptoms of present-day society as the childhood ailments of an aspiring humanity, due entirely to the excessive speed at which civilization was advancing.

INTERVIEWERS

Mein Weltbild, *Amsterdam: Querido Verlag, 1934.*

To be called to account publicly for everything one has said, even in jest, in an excess of high spirits or in momentary anger, may possibly be awkward, yet up to a point it is reasonable and natural. But to be called to account publicly for what others have said in one's name, when one cannot defend oneself, is indeed a sad predicament. "But to whom does such a thing happen?" you will ask. Well, everyone who is of sufficient interest to the public to be pursued by interviewers. You smile incredulously, but I have had plenty of direct experience and will tell you about it.

Imagine the following situation. One morning a reporter comes to you and asks you in a friendly way to tell him something about your friend N. At first you no doubt feel something approaching indignation at such a proposal. But you soon discover that there is no escape. If you refuse to say anything, the man writes: "I asked one of N's supposedly best friends about him. But he prudently avoided my questions. This in itself enables the reader to draw the inevitable conclusions." There is, therefore, no escape, and you give the following information: "Mr. N is a cheerful, straightforward man, much liked by all his friends. He can find a bright side to any situation. His enterprise and industry know no bounds; his job takes up his entire energies. He is devoted to his family and lays everything he possesses at his wife's feet. . . ."

Now for the reporter's version: "Mr. N takes nothing very seriously and has a gift for making himself liked, particularly as he carefully cultivates a hearty and ingratiating manner. He is so completely a slave to his job that he has no time for the considerations of any non-personal subject or for any extracurricular mental activity. He spoils his wife unbelievably and is utterly under her thumb. . . ."

A real reporter would make it much more spicy, but I expect this will be enough for you and your friend N. He reads the above, and some more like it, in the paper next morning, and his rage against you knows no bounds, however cheerful and benevolent his natural disposition may be. The injury done to him gives you untold pain, especially as you are really fond of him.

What's your next step, my friend? If you know, tell me quickly so that I may adopt your method with all speed.

CONGRATULATIONS TO A CRITIC

Mein Weltbild, *Amsterdam: Querido Verlag, 1934.*

To see with one's own eyes, to feel and judge without succumbing to the suggestive power of the fashion of the day, to be able to express what one has seen and felt in a trim sentence or even in a cunningly wrought word—is that not glorious? Is it not a proper subject for congratulation?

TO THE SCHOOLCHILDREN OF JAPAN

Einstein visited Japan in 1922. This message published in Mein Weltbild, *Amsterdam: Querido Verlag, 1934.*

In sending this greeting to you Japanese schoolchildren, I can lay claim to a special right to do so. For I have myself visited your beautiful country, seen its cities and houses, its mountains and woods, and the Japanese boys who had learned to love their country for its beauty. A big fat book full of colored drawings by Japanese children lies always on my table.

If you get my message of greeting from all this distance, remember that ours is the first age in history to bring about friendly and understanding intercourse between people of different nationalities; in former times nations passed their lives in mutual ignorance, and in fact in hatred or fear of one another. May the spirit of brotherly understanding gain more and more ground among them. With this in mind I, an old man, greet you Japanese schoolchildren from afar and hope that your generation may some day put mine to shame.

MESSAGE IN THE TIME-CAPSULE

World's Fair, 1939.

Our time is rich in inventive minds, the inventions of which could facilitate our lives considerably. We are crossing the seas by power and utilize power also in order to relieve humanity from all tiring muscular work. We have learned to fly and we are able to send messages and news without any difficulty over the entire world through electric waves.

However, the production and distribution of commodities is entirely unorganized so that everybody must live in fear of being eliminated from the economic cycle, in this way suffering for the want of everything. Furthermore, people living in different countries kill each other at irregular time intervals, so that also for this reason anyone who thinks about the future must live in fear and terror. This is due to the fact that the intelligence and character of the masses are incomparably lower than the intelligence and character of the few who produce something valuable for the community.

I trust that posterity will read these statements with a feeling of proud and justified superiority.

REMARKS ON BERTRAND RUSSELL'S THEORY OF KNOWLEDGE

From The Philosophy of Bertrand Russell, *Vol. V of "The Library of Living Philosophers," edited by Paul Arthur Schilpp, 1944. Translated from the original German by Paul Arthur Schilpp. Tudor Publishers.*

When the editor asked me to write something about Bertrand Russell, my admiration and respect for that author at once induced me to say yes. I owe innumerable happy hours to the reading of Russell's works, something which I cannot say of any other contemporary scientific writer, with the exception of Thorstein Veblen. Soon, however, I discovered that it is easier to give such a promise than to fulfill it. I had promised to say something about Russell as philosopher and epistemologist. After having in full confidence begun with it, I quickly recognized what a slippery field I had ventured upon, having, due to lack of experience, until now cautiously limited myself to the field of physics. The present difficulties of his science force the physicist to come to grips with philosophical problems to a greater degree than was the case with earlier generations. Although I shall not speak here of those difficulties, it was my concern with them, more than anything else, which led me to the position outlined in this essay.

In the evolution of philosophic thought through the centuries the following question has played a major rôle: what knowledge is pure thought able to supply independently of sense perception? Is there any such knowledge? If not, what precisely is the relation between our knowledge and the raw material furnished by sense impressions? An almost boundless chaos of philosophical opinions corresponds to these questions and to a few others intimately connected with them. Nevertheless there is visible in this process of relatively fruitless but heroic endeavors a systematic trend of development, namely, an increasing skepticism concerning every attempt by means of pure thought to learn something about the "objective world," about the world of "things" in contrast to the world of mere "concepts and ideas." Be it said parenthetically that, just as on the part of a real philosopher, quotation marks are used here to introduce an illegitimate concept, which the reader is asked to permit for the moment, although the concept is suspect in the eyes of the philosophical police.

During philosophy's childhood it was rather generally believed that it is possible to find everything which can be known by means of mere reflection. It was an illusion which anyone can easily understand if, for a moment, he dismisses what he has learned from later philosophy and from natural science; he will not be surprised to find that Plato ascribed a higher reality to "ideas" than to empirically experienceable things. Even in Spinoza and as late as in Hegel this prejudice was the vitalizing force which seems still to have played the major rôle. Someone, indeed, might even raise the question whether, without something of this illusion, anything really great can be achieved in the realm of philosophic thought—but we do not wish to ask this question.

This more aristocratic illusion concerning the unlimited penetrative power of thought has as its counterpart the more plebeian illusion of naïve realism, according to which things "are" as they are perceived by us through our senses. This illusion dominates the daily life of men and of animals; it is also the point of departure in all of the sciences, especially of the natural sciences.

These two illusions cannot be overcome independently. The overcoming of naïve realism has been relatively simple. In his introduction to his volume, *An Inquiry Into Meaning and Truth,* Russell has characterized this process in a marvelously concise fashion:

> We all start from "naïve realism," i.e., the doctrine that things are what they seem. We think that grass is green, that stones are hard, and that snow is cold. But physics assures us that the greenness of grass, the hardness of stones, and the coldness of snow are not the greenness, hardness, and coldness that we know in our own experience, but something very different. The observer, when he seems to himself to be observing a stone, is really, if physics is to be believed, observing the effects of the stone upon himself. Thus science seems to be at war with itself: when it most means to be objective, it finds itself plunged into subjectivity against its will. Naïve realism leads to physics, and physics, if true, shows that naïve realism is false. Therefore naïve realism, if true, is false; therefore it is false. (pp. 14–15)

Apart from their masterful formulation these lines say something which had never previously occurred to me. For, superficially considered, the mode of thought in Berkeley and Hume seems to stand in contrast to the mode of thought in the natural sciences. However, Russell's just cited remark uncovers a connection: if Berkeley relies upon the fact that we do not directly grasp the "things" of the external world through our senses, but that only events causally connected with the presence of "things" reach our sense organs, then this is a consideration which gets its persuasive character from our confidence in the physical mode of thought. For, if one doubts the physical mode of thought in even its most general features, there is no necessity to interpolate between the object and the act of vision anything which separates the object from the subject and makes the "existence of the object" problematical.

It was, however, the very same physical mode of thought and its practical successes which have shaken the confidence in the possibility of understanding things and their relations by means of purely speculative thought. Gradually the conviction gained recognition that all knowledge about things is exclusively a working-over of the raw material furnished by the senses. In this general (and intentionally somewhat vaguely stated) form this sentence is probably today commonly accepted. But this conviction does not rest on the supposition that anyone has actually proved the impossibility of gaining knowledge of reality by means of pure speculation, but rather upon the fact that the empirical (in the above-mentioned sense) procedure alone has shown its capacity to be the source of knowledge. Galileo and Hume first upheld this principle with full clarity and decisiveness.

Hume saw that concepts which we must regard as essential, such as, for example, causal connection, cannot be gained from material given to us by the senses. This insight led him to a skeptical attitude as concerns knowledge of any kind. If one reads Hume's books, one is amazed that many and sometimes even highly esteemed philosophers after him have been able to write so much obscure stuff and even find grateful readers for it. Hume has permanently influenced the development of the best of philosophers who came after him. One senses him in the reading of Russell's philosophical analyses, whose acumen and simplicity of expression have often reminded me of Hume.

Man has an intense desire for assured knowledge. That is why Hume's clear message seemed crushing: the sensory raw material, the only source of our knowledge, through habit may lead us to belief and expectation but not to the knowledge and still less to the understanding of lawful relations. Then Kant took the stage with an idea which, though certainly untenable in the form in which he put it, signified a step towards the solution of Hume's dilemma: whatever in knowledge is of empirical origin is never certain (Hume). If, therefore, we have definitely assured knowledge, it must be grounded in reason itself. This is held to be the case, for example, in the propositions of geometry and in the principle of causality. These and certain other types of knowledge are, so to speak, a part of the implements of thinking and therefore do not previously have to be gained from sense data (i.e., they are *a priori* knowledge). Today everyone knows, of course, that the mentioned concepts contain nothing of the certainty, of the inherent necessity, which Kant had attributed to them. The following, however, appears to me to be correct in Kant's statement of the problem: in thinking we use, with a certain "right," concepts to which there is no access from the materials of sensory experience, if the situation is viewed from the logical point of view.

As a matter of fact, I am convinced that even much more is to be asserted: the concepts which arise in our thought and in our linguistic expressions are all—when viewed logically—the free creations of thought which cannot inductively be gained from sense experiences. This is not so easily noticed only because we have the habit of combining certain concepts and conceptual relations (propositions) so definitely with certain sense experiences that we do not become conscious of the gulf—logically unbridgeable—which separates the world of sensory experiences from the world of concepts and propositions.

Thus, for example, the series of integers is obviously an invention of the human mind, a self-created tool which simplifies the ordering of certain sensory experiences. But there is no way in which this concept could be made to grow, as it were, directly out of sense experiences. It is deliberately that I choose here the concept of number, because it belongs to pre-scientific thinking and because, in spite of that fact, its constructive character is still easily recognizable. The more, however, we turn to the most primitive concepts of everyday life, the more difficult it becomes amidst the mass of inveterate habits to recognize the concept as an independent creation of thinking. It was thus that the fateful conception—fateful, that is to say, for an understanding of the here-existing conditions—could arise, according to which the concepts originate from experience by way of "abstraction," i.e., through omission of a part of its content. I want to indicate now why this conception appears to me to be so fateful.

As soon as one is at home in Hume's critique one is easily led to believe that all those concepts and propositions which cannot be deduced from the sensory raw material are, on account of their "metaphysical" character, to be removed from thinking. For all thought acquires material content only through its relationship with that sensory material. This latter proposition I take to be entirely true; but I hold the prescription for thinking which is grounded on this proposition to be false. For this claim—if only carried through consistently—absolutely excludes thinking of any kind as "metaphysical."

In order that thinking might not degenerate into "metaphysics," or into empty talk, it is only necessary that enough propositions of the conceptual system be firmly enough connected with sensory experiences and that the conceptional system, in view of its task of ordering and surveying sense experience, should show as much unity and parsimony as possible. Beyond that, however, the "system" is (as regards logic) a free play with symbols according to (logically) arbitrarily given rules of the game. All this applies as much (and in the same manner) to the thinking in daily life as to the more consciously and systematically constructed thinking in the sciences.

It will now be clear what is meant if I make the following

statement: by his clear critique Hume did not only advance philosophy in a decisive way but also—though through no fault of his—created a danger for philosophy in that, following his critique, a fateful "fear of metaphysics" arose which has come to be a malady of contemporary empiricistic philosophizing; this malady is the counterpart to that earlier philosophizing in the clouds, which thought it could neglect and dispense with what was given by the senses.

No matter how much one may admire the acute analysis which Russell has given us in his latest book on *Meaning and Truth,* it still seems to me that even there the specter of the metaphysical fear has caused some damage. For this fear seems to me, for example, to be the cause for conceiving of the "thing" as a "bundle of qualities," such that the "qualities" are to be taken from the sensory raw material. Now the fact that two things are said to be one and the same thing, if they coincide in all qualities, forces one to consider the geometrical relations between things as belonging to their qualities. (Otherwise one is forced to look upon the Eiffel Tower in Paris and a New York skyscraper as "the same thing.")* However, I see no "metaphysical" danger in taking the thing (the object in the sense of physics) as an independent concept into the system together with the proper spatio-temporal structure.

In view of these endeavors I am particularly pleased to note that, in the last chapter of the book, it finally turns out that one can, after all, not get along without "metaphysics." The only thing to which I take exception there is the bad intellectual conscience which shines through between the lines.

---

* Compare Russell's *An Inquiry Into Meaning and Truth,* 119–120, chapter on "Proper Names."

A MATHEMATICIAN'S MIND

Testimonial for An Essay on the Psychology of Invention in the Mathematical Field *by Jacques S. Hadamard, Princeton University Press, 1945.*

Jacques Hadamard, a French mathematician, conducted a psychological survey of mathematicians to determine their mental processes at work. Below are two of the questions followed by Albert Einstein's answers.

It would be very helpful for the purpose of psychological investigation to know what internal or mental images, what kind of "internal words" mathematicians make use of; whether they are motor, auditory, visual, or mixed, depending on the subject which they are studying.

Especially in research thought, do the mental pictures or internal words present themselves in the full consciousness or in the fringe-consciousness . . . ?

My dear Colleague:

In the following, I am trying to answer in brief your questions as well as I am able. I am not satisfied myself with those answers and I am willing to answer more questions if you believe this could be of any advantage for the very interesting and difficult work you have undertaken.

(A) The words or the language, as they are written or spoken, do not seem to play any rôle in my mechanism of thought. The psychical entities which seem to serve as elements in thought are certain signs and more or less clear images which can be "voluntarily" reproduced and combined.

There is, of course, a certain connection between those elements and relevant logical concepts. It is also clear that the desire to arrive finally at logically connected concepts is the emotional basis of this rather vague play with the above-mentioned elements. But taken from a psychological viewpoint, this combinatory play seems to be the essential feature in productive thought—before there is any connection with logical construction in words or other kinds of signs which can be communicated to others.

(B) The above-mentioned elements are, in my case, of visual and some of muscular type. Conventional words or other signs have to be sought for laboriously only in a secondary stage, when the mentioned associative play is sufficiently established and can be reproduced at will.

(C) According to what has been said, the play with the mentioned elements is aimed to be analogous to certain logical connections one is searching for.

(D) Visual and motor. In a stage when words intervene at all, they are, in my case, purely auditive, but they interfere only in a secondary stage, as already mentioned.

(E) It seems to me that what you call full consciousness is a limit case which can never be fully accomplished. This seems to me connected with the fact called the narrowness of consciousness (*Enge des Bewusstseins*).

Remark: Professor Max Wertheimer has tried to investigate the distinction between mere associating or combining of reproducible elements and between understanding (*organisches Begreifen*); I cannot judge how far his psychological analysis catches the essential point.

THE STATE AND THE INDIVIDUAL CONSCIENCE

An open letter to the Society for Social Responsibility in Science, published in Science, Vol. 112, December 22, 1950, p. 760.

DEAR FELLOW-SCIENTISTS:

The problem of how man should act if his government prescribes actions or society expects an attitude which his own conscience considers wrong is indeed an old one. It is easy to say that the individual cannot be held responsible for acts carried out under irresistible compulsion, because the individual is fully dependent upon the society in which he is living and therefore must accept its rules. But the very formulation of this idea makes it obvious to what extent such a concept contradicts our sense of justice.

External compulsion can, to a certain extent, reduce but never cancel the responsibility of the individual. In the Nuremberg trials this idea was considered to be self-evident. Whatever is morally important in our institutions, laws, and mores can be traced back to interpretation of the sense of justice of countless individuals. Institutions are in a moral sense impotent unless they are supported by the sense of responsibility of living individuals. An effort to arouse and strengthen this sense of responsibility of the individual is an important service to mankind.

In our times scientists and engineers carry particular moral responsibility, because the development of military means of mass destruction is within their sphere of activity. I feel, therefore, that the formation of the Society for Social Responsibility in Science satisfies a true need. This society, through discussion of the inherent problems, will make it easier for the individual to clarify his mind and arrive at a clear position as to his own stand; moreover, mutual help is essential for those who face difficulties because they follow their conscience.

APHORISMS FOR LEO BAECK

From the two-volume commemorative publication in honor of the eightieth birthday of Leo Baeck, May 23, 1953.

I salute the man who is going through life always helpful, knowing no fear, and to whom aggressiveness and resentment are alien. Such is the stuff of which the great moral leaders are made who proffer consolation to mankind in their self-created miseries.

The attempt to combine wisdom and power has only rarely been successful and then only for a short while.

Man usually avoids attributing cleverness to somebody else
—unless it is an enemy.

Few people are capable of expressing with equanimity opinions which differ from the prejudices of their social environment. Most people are even incapable of forming such opinions.

The majority of the stupid is invincible and guaranteed for
all time. The terror of their tyranny, however, is alleviated by
their lack of consistency.

In order to form an immaculate member of a flock of sheep
one must, above all, be a sheep.

The contrasts and contradictions that can permanently live
peacefully side by side in a skull make all the systems of political
optimists and pessimists illusory.

Whoever undertakes to set himself up as judge in the field of
Truth and Knowledge is shipwrecked by the laughter of the
gods.

Joy in looking and comprehending is nature's most beautiful gift.

About Freedom

ON ACADEMIC FREEDOM

*Apropos of the Gumbel case, 1931. E. J. Gumbel, professor
at the University of Heidelberg, Germany, had coura-
geously exposed political assassinations by German Nazis
and other members of the extreme right. In consequence he
was violently attacked, particularly by right-wing students.
Published in* Mein Weltbild, *Amsterdam: Querido Verlag,
1934.*

Numerous are the academic chairs, but rare are wise and
noble teachers. Numerous and large are the lecture halls, but

far from numerous the young people who genuinely thirst for truth and justice. Numerous are the wares that nature produces by the dozen, but her choice products are few.

We all know that, so why complain? Was it not always thus and will it not always thus remain? Certainly, and one must take what nature gives as one finds it. But there is also such a thing as a spirit of the times, an attitude of mind characteristic of a particular generation, which is passed on from individual to individual and gives its distinctive mark to a society. Each of us has to do his little bit toward transforming this spirit of the times.

Compare the spirit which animated the youth in our universities a hundred years ago with that prevailing today. They had faith in the amelioration of human society, respect for every honest opinion, the tolerance for which our great minds had lived and fought. In those days men strove for a larger political unity, which at that time was called Germany. It was the students and the teachers at the universities in whom these ideals were alive.

Today also there is an urge toward social progress, toward tolerance and freedom of thought, toward a larger political unity, which we today call Europe. But the students at our universities have ceased as completely as their teachers to embody the hopes and ideals of the people. Anyone who looks at our times soberly and dispassionately must admit this.

We are assembled today to take stock of ourselves. The external reason for this meeting is the Gumbel case. This apostle of justice has written about unexpiated political crimes with devoted industry, high courage, and exemplary fairness, and has done the community a signal service by his books. And this is the man whom the students and a good many of the faculty of his university are today doing their best to expel.

Political passion cannot be allowed to go to such lengths. I am convinced that every man who reads Mr. Gumbel's books with an open mind will get the same impression from them as I have. Men like him are needed if we are ever to build up a healthy political society.

Let every man judge by himself, by what he has himself read, not by what others tell him.

If that happens, this Gumbel case, after an unedifying beginning, may still do good.

FASCISM AND SCIENCE

A letter to Signor Rocco, Minister of Justice and Education under Mussolini, 1925–1932. Published in Mein Weltbild, *Amsterdam: Querido Verlag, 1934.*

MY DEAR SIR:

Two of the most eminent and respected men of science in Italy have applied to me in their difficulties of conscience and requested me to write to you with the object of preventing, if possible, a cruel hardship with which men of learning are threatened in Italy. I refer to an oath of loyalty to the Fascist system. The burden of my request is that you should please advise Signor Mussolini to spare the flower of Italy's intellect this humiliation.

However much our political convictions may differ, I know that we agree on one basic point: we both admire the outstanding accomplishments of the European intellect and see in them our highest values. Those achievements are based on the freedom of thought and of teaching, on the principle that the desire for truth must take precedence over all other desires. It was this basis alone that enabled our civilization to take its rise in Greece and to celebrate its rebirth in Italy at the Renaissance. This, our most valuable possession, has been paid for by the martyr's blood of pure and great men, for whose sake Italy is still loved and revered today.

I do not intend to argue with you about what inroads on human liberty may be justified by reasons of state. But the pursuit of scientific truth, detached from the practical interests of everyday life, ought to be treated as sacred by every government, and it is in the highest interests of all that honest servants of truth should be left in peace. This is also undoubtedly in the interests of the Italian state and its prestige in the eyes of the world.

ON FREEDOM

From Freedom, Its Meaning, *edited by Ruth Nanda Anshen, New York: Harcourt, Brace and Company, 1940. Translated by James Gutmann.*

I know that it is a hopeless undertaking to debate about fundamental value judgments. For instance, if someone approves, as a goal, the extirpation of the human race from the earth, one cannot refute such a viewpoint on rational grounds. But if there is agreement on certain goals and values, one can argue rationally about the means by which these objectives may be attained. Let us, then, indicate two goals which may well be agreed upon by nearly all who read these lines.

1. Those instrumental goods which should serve to maintain the life and health of all human beings should be produced by the least possible labor of all.

2. The satisfaction of physical needs is indeed the indispensable precondition of a satisfactory existence, but in itself it is not enough. In order to be content, men must also have the possibility of developing their intellectual and artistic powers to whatever extent accords with their personal characteristics and abilities.

The first of these two goals requires the promotion of all knowledge relating to the laws of nature and the laws of social processes, that is, the promotion of all scientific endeavor. For scientific endeavor is a natural whole, the parts of which mutually support one another in a way which, to be sure, no one can anticipate. However, the progress of science presupposes the possibility of unrestricted communication of all results and judgments—freedom of expression and instruction in all realms of intellectual endeavor. By freedom I understand social conditions of such a kind that the expression of opinions and assertions about general and particular matters of knowledge will not involve dangers or serious disadvantages for him who expresses them. This freedom of communication is indispensable for the development and extension of scientific knowledge, a consideration of much practical import. In the first instance it must be guaranteed by law. But laws alone cannot secure freedom of expression; in order that every man may present his views without penalty, there must be a spirit of tolerance in the entire population. Such an ideal of external liberty can never be fully attained but must be sought unremittingly if scientific thought, and philosophical and creative thinking in general, are to be advanced as far as possible.

If the second goal, that is, the possibility of the spiritual development of all individuals, is to be secured, a second kind of outward freedom is necessary. Man should not have to work for the achievement of the necessities of life to such an extent that he has neither time nor strength for personal activities. Without this second kind of outward liberty, freedom of expression is useless for him. Advances in technology would provide the possibility of this kind of freedom if the problem of a reasonable division of labor were solved.

The development of science and of the creative activities of the spirit in general requires still another kind of freedom, which may be characterized as inward freedom. It is this freedom of the spirit which consists in the independence of thought from the restrictions of authoritarian and social prejudices as well as from unphilosophical routinizing and habit in general. This inward freedom is an infrequent gift of nature and a worthy objective for the individual. Yet the community can do much to further this achievement, too, at least by not interfering with its development. Thus schools may interfere with the development of inward freedom through authoritarian influences and through imposing on young people excessive spiritual burdens; on the other hand, schools may favor such freedom by encouraging independent thought. Only if outward and inner freedom are constantly and consciously pursued is there a possibility of spiritual development and perfection and thus of improving man's outward and inner life.

ADDRESS ON RECEIVING
LORD & TAYLOR AWARD

Broadcast by radio (tape-recorded) May 4, 1953.

I gladly accept this award as an expression of friendly sentiments. It gives me great pleasure, indeed, to see the stubbornness of an incorrigible nonconformist warmly acclaimed. To be sure, we are concerned here with nonconformism in a remote field of endeavor, and no Senatorial committee has as yet felt impelled to tackle the important task of combating, also in this field, the dangers which threaten the inner security of the uncritical or else intimidated citizen.

As for the words of warm praise addressed to me, I shall carefully refrain from disputing them. For who still believes that there is such a thing as genuine modesty? I should run the risk of being taken for just an old hypocrite. You will surely understand that I do not find the courage to brave this danger.

Thus all that remains is to assure you of my gratitude.

MODERN INQUISITIONAL METHODS

Letter to William Frauenglass, a teacher in Brooklyn, N. Y., who had refused to testify before a Congressional Committee. Published June 12, 1953, in the New York Times.

May 16, 1953

Dear Mr. Frauenglass:

Thank you for your communication. By "remote field" I referred to the theoretical foundations of physics.

The problem with which the intellectuals of this country are confronted is very serious. The reactionary politicians have managed to instill suspicion of all intellectual efforts into the public by dangling before their eyes a danger from without.

Having succeeded so far, they are now proceeding to suppress the freedom of teaching and to deprive of their positions all those who do not prove submissive, i.e., to starve them.

What ought the minority of intellectuals to do against this evil? Frankly, I can only see the revolutionary way of non-co-operation in the sense of Gandhi's. Every intellectual who is called before one of the committees ought to refuse to testify, i.e., he must be prepared for jail and economic ruin, in short, for the sacrifice of his personal welfare in the interest of the cultural welfare of his country.

However, this refusal to testify must not be based on the well-known subterfuge of invoking the Fifth Amendment against possible self-incrimination, but on the assertion that it is shameful for a blameless citizen to submit to such an inquisition and that this kind of inquisition violates the spirit of the Constitution.

If enough people are ready to take this grave step they will be successful. If not, then the intellectuals of this country deserve nothing better than the slavery which is intended for them.

P.S. This letter need not be considered "confidential."

HUMAN RIGHTS

Address to Chicago Decalogue Society, February 20, 1954.

LADIES AND GENTLEMEN:

You are assembled today to devote your attention to the problem of human rights. You have decided to offer me an award on this occasion. When I learned about it, I was some-what depressed by your decision. For in how unfortunate a state must a community find itself if it cannot produce a more suitable candidate upon whom to confer such a distinction?

In a long life I have devoted all my faculties to reach a some-what deeper insight into the structure of physical reality. Never have I made any systematic effort to ameliorate the lot of men, to fight injustice and suppression, and to improve the traditional forms of human relations. The only thing I did was this: in long intervals I have expressed an opinion on public issues whenever they appeared to me so bad and unfortunate that silence would have made me feel guilty of complicity.

The existence and validity of human rights are not written in the stars. The ideals concerning the conduct of men toward each other and the desirable structure of the community have been conceived and taught by enlightened individuals in the course of history. Those ideals and convictions which resulted from historical experience, from the craving for beauty and harmony, have been readily accepted in theory by man—and at all times, have been trampled upon by the same people under the pressure of their animal instincts. A large part of history is therefore replete with the struggle for those human rights, an eternal struggle in which a final victory can never be won. But to tire in that struggle would mean the ruin of society.

In talking about human rights today, we are referring primarily to the following demands: protection of the individual against arbitrary infringement by other individuals or by the government; the right to work and to adequate earnings from work; freedom of discussion and teaching; adequate participation of the individual in the formation of his government. These human rights are nowadays recognized theoretically, although, by abundant use of formalistic, legal maneuvers, they are being violated to a much greater extent than even a generation ago. There is, however, one other human right which is infrequently mentioned but which seems to be destined to become very important: this is the right, or the duty, of the individual to abstain from cooperating in activities which he considers wrong or pernicious. The first place in this respect must be given to the refusal of military service. I have known instances where individuals of unusual moral strength and integrity have, for that reason, come into conflict with the organs of the state. The Nuremberg Trial of the German war criminals was tacitly based on the recognition of the principle: criminal actions cannot be excused if committed on government orders; conscience supersedes the authority of the law of the state.

The struggle of our own days is being waged primarily for the freedom of political conviction and discussion as well as for the freedom of research and teaching. The fear of Communism has led to practices which have become incomprehensible to the rest of civilized mankind and exposed our country to ridicule. How long shall we tolerate that politicians, hungry for power, try to gain political advantages in such a way? Sometimes it seems that people have lost their sense of humor to such a degree that the French saying, "Ridicule kills," has lost its validity.

About Religion

RELIGION AND SCIENCE

Written expressly for the New York Times Magazine. *Appeared there November 9, 1930 (pp. 1-4). The German text was published in the* Berliner Tageblatt, *November 11, 1930.*

Everything that the human race has done and thought is concerned with the satisfaction of deeply felt needs and the assuagement of pain. One has to keep this constantly in mind if one wishes to understand spiritual movements and their development. Feeling and longing are the motive force behind all human endeavor and human creation, in however exalted a guise the latter may present themselves to us. Now what are the feelings and needs that have led men to religious thought and belief in the widest sense of the words? A little consideration will suffice to show us that the most varying emotions preside over the birth of religious thought and experience. With primitive man it is above all fear that evokes religious notions— fear of hunger, wild beasts, sickness, death. Since at this stage of existence understanding of causal connections is usually poorly developed, the human mind creates illusory beings more or less analogous to itself on whose wills and actions these fearful happenings depend. Thus one tries to secure the favor of these beings by carrying out actions and offering sacrifices which, according to the tradition handed down from generation to generation, propitiate them or make them well disposed toward a mortal. In this sense I am speaking of a religion of fear. This, though not created, is in an important degree stabilized by the formation of a special priestly caste which sets itself up as a mediator between the people and the beings they fear, and erects a hegemony on this basis. In many cases a leader or ruler or a privileged class whose position rests on other factors combines priestly functions with its secular authority in order to make the latter more secure; or the political rulers and the priestly caste make common cause in their own interests.

The social impulses are another source of the crystallization of religion. Fathers and mothers and the leaders of larger human communities are mortal and fallible. The desire for guidance, love, and support prompts men to form the social or moral conception of God. This is the God of Providence, who protects, disposes, rewards, and punishes; the God who, according to the limits of the believer's outlook, loves and cherishes the life of the tribe or of the human race, or even life itself; the comforter in sorrow and unsatisfied longing; he who preserves the souls of the dead. This is the social or moral conception of God.

The Jewish scriptures admirably illustrate the development from the religion of fear to moral religion, a development continued in the New Testament. The religions of all civilized peoples, especially the peoples of the Orient, are primarily moral religions. The development from a religion of fear to moral religion is a great step in peoples' lives. And yet, that primitive religions are based entirely on fear and the religions of civilized peoples purely on morality is a prejudice against which we must be on our guard. The truth is that all religions are a varying blend of both types, with this differentiation: that on the higher levels of social life the religion of morality predominates.

Common to all these types is the anthropomorphic character of their conception of God. In general, only individuals of exceptional endowments, and exceptionally high-minded communities, rise to any considerable extent above this level. But there is a third stage of religious experience which belongs to all of them, even though it is rarely found in a pure form: I shall call it cosmic religious feeling. It is very difficult to elucidate this feeling to anyone who is entirely without it, especially as there is no anthropomorphic conception of God corresponding to it.

The individual feels the futility of human desires and aims and the sublimity and marvelous order which reveal themselves both in nature and in the world of thought. Individual existence impresses him as a sort of prison and he wants to experience the universe as a single significant whole. The beginnings of cosmic religious feeling already appear at an early stage of development, e.g., in many of the Psalms of David and in some of the Prophets. Buddhism, as we have learned especially from the wonderful writings of Schopenhauer, contains a much stronger element of this.

The religious geniuses of all ages have been distinguished by this kind of religious feeling, which knows no dogma and no God conceived in man's image; so that there can be no church whose central teachings are based on it. Hence it is precisely among the heretics of every age that we find men who were filled with this highest kind of religious feeling and were in many cases regarded by their contemporaries as atheists, sometimes also as saints. Looked at in this light, men like Democritus, Francis of Assisi, and Spinoza are closely akin to one another.

How can cosmic religious feeling be communicated from one person to another, if it can give rise to no definite notion of a God and no theology? In my view, it is the most important function of art and science to awaken this feeling and keep it alive in those who are receptive to it.

We thus arrive at a conception of the relation of science to religion very different from the usual one. When one views the matter historically, one is inclined to look upon science and religion as irreconcilable antagonists, and for a very obvious reason. The man who is thoroughly convinced of the universal operation of the law of causation cannot for a moment entertain the idea of a being who interferes in the course of events— provided, of course, that he takes the hypothesis of causality really seriously. He has no use for the religion of fear and equally little for social or moral religion. A God who rewards and punishes is inconceivable to him for the simple reason that a man's actions are determined by necessity, external and internal, so that in God's eyes he cannot be responsible, any more than an inanimate object is responsible for the motions it undergoes. Science has therefore been charged with undermining morality, but the charge is unjust. A man's ethical behavior should be based effectually on sympathy, education, and social ties and needs; no religious basis is necessary. Man would indeed be in a poor way if he had to be restrained by fear of punishment and hope of reward after death.

It is therefore easy to see why the churches have always fought science and persecuted its devotees. On the other hand, I maintain that the cosmic religious feeling is the strongest and noblest motive for scientific research. Only those who realize the immense efforts and, above all, the devotion without which pioneer work in theoretical science cannot be achieved are able to grasp the strength of the emotion out of which alone such work, remote as it is from the immediate realities of life, can issue. What a deep conviction of the rationality of the universe and what a yearning to understand, were it but a feeble reflection of the mind revealed in this world, Kepler and Newton must have had to enable them to spend years of solitary labor in disentangling the principles of celestial mechanics! Those whose acquaintance with scientific research is derived chiefly from its practical results easily develop a completely false notion of the mentality of the men who, surrounded by a skeptical world, have shown the way to kindred spirits scattered wide

through the world and the centuries. Only one who has devoted his life to similar ends can have a vivid realization of
what has inspired these men and given them the strength to
remain true to their purpose in spite of countless failures. It
is cosmic religious feeling that gives a man such strength. A
contemporary has said, not unjustly, that in this materialistic
age of ours the serious scientific workers are the only profoundly
religious people.

THE RELIGIOUS SPIRIT OF SCIENCE

Mein Weltbild, *Amsterdam: Querido Verlag, 1934.*

You will hardly find one among the profounder sort of
scientific minds without a religious feeling of his own. But it
is different from the religiosity of the naïve man. For the latter,
God is a being from whose care one hopes to benefit and whose
punishment one fears; a sublimation of a feeling similar to that
of a child for its father, a being to whom one stands, so to
speak, in a personal relation, however deeply it may be tinged
with awe.

But the scientist is possessed by the sense of universal causation. The future, to him, is every whit as necessary and determined as the past. There is nothing divine about morality; it
is a purely human affair. His religious feeling takes the form of
a rapturous amazement at the harmony of natural law, which
reveals an intelligence of such superiority that, compared with
it, all the systematic thinking and acting of human beings is
an utterly insignificant reflection. This feeling is the guiding
principle of his life and work, in so far as he succeeds in keeping
himself from the shackles of selfish desire. It is beyond question
closely akin to that which has possessed the religious geniuses of
all ages.

SCIENCE AND RELIGION

*Part I from an address at Princeton Theological Seminary,
May 19, 1939; published in* Out of My Later Years, *New
York: Philosophical Library, 1950. Part II from* Science,
Philosophy and Religion, *A Symposium, published by the
Conference on Science, Philosophy and Religion in Their
Relation to the Democratic Way of Life, Inc., New York,
1941.*

I.

During the last century, and part of the one before, it was
widely held that there was an unreconcilable conflict between
knowledge and belief. The opinion prevailed among advanced
minds that it was time that belief should be replaced increasingly by knowledge; belief that did not itself rest on knowledge
was superstition, and as such had to be opposed. According
to this conception, the sole function of education was to open
the way to thinking and knowing, and the school, as the outstanding organ for the people's education, must serve that end
exclusively.

One will probably find but rarely, if at all, the rationalistic
standpoint expressed in such crass form; for any sensible man
would see at once how one-sided is such a statement of the
position. But it is just as well to state a thesis starkly and
nakedly, if one wants to clear up one's mind as to its nature.

It is true that convictions can best be supported with experience and clear thinking. On this point one must agree unre-
servedly with the extreme rationalist. The weak point of his
conception is, however, this, that those convictions which are
necessary and determinant for our conduct and judgments
cannot be found solely along this solid scientific way.

For the scientific method can teach us nothing else beyond
how facts are related to, and conditioned by, each other. The
aspiration toward such objective knowledge belongs to the

highest of which man is capable, and you will certainly not
suspect me of wishing to belittle the achievements and the
heroic efforts of man in this sphere. Yet it is equally clear that
knowledge of what *is* does not open the door directly to what
should be. One can have the clearest and most complete
knowledge of what *is,* and yet not be able to deduct from that
what should be the *goal* of our human aspirations. Objective
knowledge provides us with powerful instruments for the
achievements of certain ends, but the ultimate goal itself and
the longing to reach it must come from another source. And
it is hardly necessary to argue for the view that our existence
and our activity acquire meaning only by the setting up of
such a goal and of corresponding values. The knowledge of
truth as such is wonderful, but it is so little capable of acting
as a guide that it cannot prove even the justification and the
value of the aspiration toward that very knowledge of truth.
Here we face, therefore, the limits of the purely rational conception of our existence.

But it must not be assumed that intelligent thinking can
play no part in the formation of the goal and of ethical judgments. When someone realizes that for the achievement of
an end certain means would be useful, the means itself
becomes thereby an end. Intelligence makes clear to us the
interrelation of means and ends. But mere thinking cannot
give us a sense of the ultimate and fundamental ends. To
make clear these fundamental ends and valuations, and to
set them fast in the emotional life of the individual, seems to
me precisely the most important function which religion has
to perform in the social life of man. And if one asks whence
derives the authority of such fundamental ends, since they
cannot be stated and justified merely by reason, one can only
answer: they exist in a healthy society as powerful traditions,
which act upon the conduct and aspirations and judgments
of the individuals; they are there, that is, as something living,
without its being necessary to find justification for their existence. They come into being not through demonstration but

through revelation, through the medium of powerful personalities. One must not attempt to justify them, but rather to sense their nature simply and clearly.

The highest principles for our aspirations and judgments are given to us in the Jewish-Christian religious tradition. It is a very high goal which, with our weak powers, we can reach only very inadequately, but which gives a sure foundation to our aspirations and valuations. If one were to take that goal out of its religious form and look merely at its purely human side, one might state it perhaps thus: free and responsible development of the individual, so that he may place his powers freely and gladly in the service of all mankind.

There is no room in this for the divinization of a nation, of a class, let alone of an individual. Are we not all children of one father, as it is said in religious language? Indeed, even the divinization of humanity, as an abstract totality, would not be in the spirit of that ideal. It is only to the individual that a soul is given. And the high destiny of the individual is to serve rather than to rule, or to impose himself in any other way.

If one looks at the substance rather than at the form, then one can take these words as expressing also the fundamental democratic position. The true democrat can worship his nation as little as can the man who is religious, in our sense of the term.

What, then, in all this, is the function of education and of the school? They should help the young person to grow up in such a spirit that these fundamental principles should be to him as the air which he breathes. Teaching alone cannot do that.

If one holds these high principles clearly before one's eyes, and compares them with the life and spirit of our times, then it appears glaringly that civilized mankind finds itself at present in grave danger. In the totalitarian states it is the rulers themselves who strive actually to destroy that spirit of humanity. In less threatened parts it is nationalism and intolerance, as well as the oppression of the individuals by economic means, which threaten to choke these most precious traditions.

A realization of how great is the danger is spreading, however, among thinking people, and there is much search for means with which to meet the danger—means in the field of national and international politics, of legislation, or organization in general. Such efforts are, no doubt, greatly needed. Yet the ancients knew something which we seem to have forgotten. All means prove but a blunt instrument, if they have not behind them a living spirit. But if the longing for the achievement of the goal is powerfully alive within us, then shall we not lack the strength to find the means for reaching the goal and for translating it into deeds.

II.

It would not be difficult to come to an agreement as to what we understand by science. Science is the century-old endeavor to bring together by means of systematic thought the perceptible phenomena of this world into as thoroughgoing an association as possible. To put it boldly, it is the attempt at the posterior reconstruction of existence by the process of conceptualization. But when asking myself what religion is I cannot think of the answer so easily. And even after finding an answer which may satisfy me at this particular moment, I still remain convinced that I can never under any circumstances bring together, even to a slight extent, the thoughts of all those who have given this question serious consideration.

At first, then, instead of asking what religion is I should prefer to ask what characterizes the aspirations of a person who gives me the impression of being religious: a person who is religiously enlightened appears to me to be one who has, to the best of his ability, liberated himself from the fetters of his selfish desires and is preoccupied with thoughts, feelings, and aspirations to which he clings because of their superpersonal value. It seems to me that what is important is the force of this superpersonal content and the depth of the conviction concerning its overpowering meaningfulness, regardless of whether any attempt is made to unite this content with a divine Being, for otherwise it would not be possible to count Buddha and Spinoza as religious personalities. Accordingly, a religious person is devout in the sense that he has no doubt of the significance and loftiness of those superpersonal objects and goals which neither require nor are capable of rational foundation. They exist with the same necessity and matter-of-factness as he himself. In this sense religion is the age-old endeavor of mankind to become clearly and completely conscious of these values and goals and constantly to strengthen and extend their effect. If one conceives of religion and science according to these definitions then a conflict between them appears impossible. For science can only ascertain what *is*, but not what *should be*, and outside of its domain value judgments of all kinds remain necessary. Religion, on the other hand, deals only with evaluations of human thought and action: it cannot justifiably speak of facts and relationships between facts. According to this interpretation the well-known conflicts between religion and science in the past must all be ascribed to a misapprehension of the situation which has been described.

For example, a conflict arises when a religious community insists on the absolute truthfulness of all statements recorded in the Bible. This means an intervention on the part of religion into the sphere of science; this is where the struggle of the Church against the doctrines of Galileo and Darwin belongs. On the other hand, representatives of science have often made an attempt to arrive at fundamental judgments with respect to values and ends on the basis of scientific method, and in this way have set themselves in opposition to religion. These conflicts have all sprung from fatal errors.

Now, even though the realms of religion and science in themselves are clearly marked off from each other, nevertheless there exist between the two strong reciprocal relationships and dependencies. Though religion may be that which determines the goal, it has, nevertheless, learned from science, in the broadest sense, what means will contribute to the attainment of the goals it has set up. But science can only be created by those who are thoroughly imbued with the aspiration toward truth and understanding. This source of feeling, however, springs from the sphere of religion. To this there also belongs the faith in the possibility that the regulations valid for the world of existence are rational, that is, comprehensible to reason. I cannot conceive of a genuine scientist without that profound faith. The situation may be expressed by an image: science without religion is lame, religion without science is blind.

Though I have asserted above that in truth a legitimate conflict between religion and science cannot exist, I must nevertheless qualify this assertion once again on an essential point, with reference to the actual content of historical religions. This qualification has to do with the concept of God. During the youthful period of mankind's spiritual evolution human fantasy created gods in man's own image, who, by the operations of their will were supposed to determine, or at any rate to influence, the phenomenal world. Man sought to alter the disposition of these gods in his own favor by means of magic and prayer. The idea of God in the religions taught at present is a sublimation of that old concept of the gods. Its anthropomorphic character is shown, for instance, by the fact that men appeal to the Divine Being in prayers and plead for the fulfillment of their wishes.

Nobody, certainly, will deny that the idea of the existence of an omnipotent, just, and omnibeneficent personal God is able to accord man solace, help, and guidance; also, by virtue of its simplicity it is accessible to the most undeveloped mind. But, on the other hand, there are decisive weaknesses attached to this idea in itself, which have been painfully felt since the beginning of history. That is, if this being is omnipotent, then every occurrence, including every human action, every human thought, and every human feeling and aspiration is also His work; how is it possible to think of holding men responsible for their deeds and thoughts before such an almighty Being? In giving out punishment and rewards He would to a certain extent be passing judgment on Himself. How can this be combined with the goodness and righteousness ascribed to Him?

The main source of the present-day conflicts between the spheres of religion and of science lies in this concept of a personal God. It is the aim of science to establish general rules which determine the reciprocal connection of objects and events in time and space. For these rules, or laws of nature, absolutely general validity is required—not proven. It is mainly a program, and faith in the possibility of its accomplishment in principle is only founded on partial successes. But hardly anyone could be found who would deny these partial successes and ascribe them to human self-deception. The fact that on the basis of such laws we are able to predict the temporal behavior of phenomena in certain domains with great precision and certainty is deeply embedded in the consciousness of the modern man, even though he may have grasped very little of the contents of those laws. He need only consider that planetary courses within the solar system may be calculated in advance with great exactitude on the basis of a limited number of simple laws. In a similar way, though not with the same precision, it is possible to calculate in advance the mode of operation of an electric motor, a transmission system, or of a wireless apparatus, even when dealing with a novel development.

To be sure, when the number of factors coming into play in a phenomenological complex is too large, scientific method in most cases fails us. One need only think of the weather, in which case prediction even for a few days ahead is impossible. Nevertheless no one doubts that we are confronted with a causal connection whose causal components are in the main known to us. Occurrences in this domain are beyond the reach of exact prediction because of the variety of factors in operation, not because of any lack of order in nature.

We have penetrated far less deeply into the regularities obtaining within the realm of living things, but deeply enough nevertheless to sense at least the rule of fixed necessity. One need only think of the systematic order in heredity, and in the effect of poisons, as for instance alcohol, on the behavior of organic beings. What is still lacking here is a grasp of connections of profound generality, but not a knowledge of order in itself.

The more a man is imbued with the ordered regularity of all events the firmer becomes his conviction that there is no room left by the side of this ordered regularity for causes of a different nature. For him neither the rule of human nor the rule of divine will exists as an independent cause of natural events. To be sure, the doctrine of a personal God interfering with natural events could never be *refuted*, in the real sense, by science, for this doctrine can always take refuge in those domains in which scientific knowledge has not yet been able to set foot.

But I am persuaded that such behavior on the part of the representatives of religion would not only be unworthy but also fatal. For a doctrine which is able to maintain itself not in clear light but only in the dark, will of necessity lose its effect on mankind, with incalculable harm to human progress. In their struggle for the ethical good, teachers of religion must have the stature to give up the doctrine of a personal God, that is, give up that source of fear and hope which in the past placed such vast power in the hands of priests. In their labors they will have to avail themselves of those forces which are capable of cultivating the Good, the True, and the Beautiful in humanity itself. This is, to be sure, a more difficult but an incomparably more worthy task.* After religious teachers accomplish the refining process indicated they will surely recognize with joy that true religion has been ennobled and made more profound by scientific knowledge.

If it is one of the goals of religion to liberate mankind as far as possible from the bondage of egocentric cravings, desires, and fears, scientific reasoning can aid religion in yet another sense. Although it is true that it is the goal of science to discover rules which permit the association and foretelling

---

* This thought is convincingly presented in Herbert Samuel's book, *Belief and Action.*

of facts, this is not its only aim. It also seeks to reduce the connections discovered to the smallest possible number of mutually independent conceptual elements. It is in this striving after the rational unification of the manifold that it encounters its greatest successes, even though it is precisely this attempt which causes it to run the greatest risk of falling a prey to illusions. But whoever has undergone the intense experience of successful advances made in this domain is moved by profound reverence for the rationality made manifest in existence. By way of the understanding he achieves a far-reaching emancipation from the shackles of personal hopes and desires, and thereby attains that humble attitude of mind toward the grandeur of reason incarnate in existence, and which, in its profoundest depths, is inaccessible to man. This attitude, however, appears to me to be religious, in the highest sense of the word. And so it seems to me that science not only purifies the religious impulse of the dross of its anthropomorphism but also contributes to a religious spiritualization of our understanding of life.

The further the spiritual evolution of mankind advances, the more certain it seems to me that the path to genuine religiosity does not lie through the fear of life, and the fear of death, and blind faith, but through striving after rational knowledge. In this sense I believe that the priest must become a teacher if he wishes to do justice to his lofty educational mission.

RELIGION AND SCIENCE: IRRECONCILABLE?

A response to a greeting sent by the Liberal Ministers' Club of New York City. Published in The Christian Register, *June, 1948.*

Does there truly exist an insuperable contradiction between religion and science? Can religion be superseded by science? The answers to these questions have, for centuries, given rise to considerable dispute and, indeed, bitter fighting. Yet, in my own mind there can be no doubt that in both cases a dispassionate consideration can only lead to a negative answer. What complicates the solution, however, is the fact that while most people readily agree on what is meant by "science," they are likely to differ on the meaning of "religion."

As to science, we may well define it for our purpose as "methodical thinking directed toward finding regulative connections between our sensual experiences." Science, in the immediate, produces knowledge and, indirectly, means of action. It leads to methodical action if definite goals are set up in advance. For the function of setting up goals and passing statements of value transcends its domain. While it is true that science, to the extent of its grasp of causative connections, may reach important conclusions as to the compatibility and incompatibility of goals and evaluations, the independent and fundamental definitions regarding goals and values remain beyond science's reach.

As regards religion, on the other hand, one is generally agreed that it deals with goals and evaluations and, in general, with the emotional foundation of human thinking and acting, as far as these are not predetermined by the inalterable hereditary disposition of the human species. Religion is concerned with man's attitude toward nature at large, with the establishing of ideals for the individual and communal life, and with mutual human relationship. These ideals religion attempts to attain by exerting an educational influence on tradition and through the development and promulgation of certain easily accessible thoughts and narratives (epics and myths) which are apt to influence evaluation and action along the lines of the accepted ideals.

It is this mythical, or rather this symbolic, content of the religious traditions which is likely to come into conflict with science. This occurs whenever this religious stock of ideas contains dogmatically fixed statements on subjects which belong in the domain of science. Thus, it is of vital importance for the preservation of true religion that such conflicts be avoided when they arise from subjects which, in fact, are not really essential for the pursuance of the religious aims.

When we consider the various existing religions as to their essential substance, that is, divested of their myths, they do not seem to me to differ as basically from each other as the proponents of the "relativistic" or conventional theory wish us to believe. And this is by no means surprising. For the moral attitudes of a people that is supported by religion need always aim at preserving and promoting the sanity and vitality of the community and its individuals, since otherwise this community is bound to perish. A people that were to honor falsehood, defamation, fraud, and murder would be unable, indeed, to subsist for very long.

When confronted with a specific case, however, it is no easy task to determine clearly what is desirable and what should be eschewed, just as we find it difficult to decide what exactly it is that makes good painting or good music. It is something that may be felt intuitively more easily than rationally comprehended. Likewise, the great moral teachers of humanity were, in a way, artistic geniuses in the art of living. In addition to the most elementary precepts directly motivated by the preservation of life and the sparing of unnecessary suffering, there are others to which, although they are apparently not quite commensurable to the basic precepts, we nevertheless attach considerable importance. Should truth, for instance, be sought unconditionally even where its attainment and its accessibility to all would entail heavy sacrifices in toil and happiness? There are many such questions which, from a rational vantage point, cannot easily be answered or cannot be answered at all. Yet, I do not think that the so-called "relativistic" viewpoint is correct, not even when dealing with the more subtle moral decisions.

When considering the actual living conditions of present-day civilized humanity from the standpoint of even the most elementary religious commands, one is bound to experience a feeling of deep and painful disappointment at what one sees. For while religion prescribes brotherly love in the relations among the individuals and groups, the actual spectacle more resembles a battlefield than an orchestra. Everywhere, in economic as well as in political life, the guiding principle is one of ruthless striving for success at the expense of one's fellowmen. This competitive spirit prevails even in school and, destroying all feelings of human fraternity and cooperation, conceives of achievement not as derived from the love for productive and thoughtful work, but as springing from personal ambition and fear of rejection.

There are pessimists who hold that such a state of affairs is necessarily inherent in human nature; it is those who propound such views that are the enemies of true religion, for they imply thereby that religious teachings are utopian ideals and unsuited to afford guidance in human affairs. The study of the social patterns in certain so-called primitive cultures, however, seems to have made it sufficiently evident that such a defeatist view is wholly unwarranted. Whoever is concerned with this problem, a crucial one in the study of religion as such, is advised to read the description of the Pueblo Indians in Ruth Benedict's book, *Patterns of Culture*. Under the hardest living conditions, this tribe has apparently accomplished the difficult task of delivering its people from the scourge of competitive spirit and of fostering in it a temperate, cooperative conduct of life, free of external pressure and without any curtailment of happiness.

The interpretation of religion, as here advanced, implies a dependence of science on the religious attitude, a relation which, in our predominantly materialistic age, is only too easily overlooked. While it is true that scientific results are entirely independent from religious or moral considerations, those individuals to whom we owe the great creative achievements of science were all of them imbued with the truly religious conviction that this universe of ours is something perfect and susceptible to the rational striving for knowledge. If this conviction had not been a strongly emotional one and if those searching for knowledge had not been inspired by Spinoza's *Amor Dei Intellectualis,* they would hardly have been capable of that untiring devotion which alone enables man to attain his greatest achievements.

THE NEED FOR ETHICAL CULTURE

Letter read on the occasion of the seventy-fifth anniversary of the Ethical Culture Society, New York, January, 1951. Published in Mein Weltbild, *Zurich: Europa Verlag, 1953.*

I feel the need of sending my congratulations and good wishes to your Ethical Culture Society on the occasion of its anniversary celebration. True, this is not a time when we can regard with satisfaction the results which honest striving on the ethical plane has achieved in these seventy-five years. For one can hardly assert that the moral aspect of human life in general is today more satisfactory than it was in 1876.

At that time the view obtained that everything was to be hoped for from enlightenment in the field of ascertainable scientific fact and from the conquest of prejudice and superstition. All this is of course important and worthy of the best efforts of the finest people. And in this regard much has been accomplished in these seventy-five years and has been disseminated by means of literature and the stage. But the clearing away of obstacles does not by itself lead to an ennoblement of social and individual life. For along with this negative result, a positive aspiration and effort for an ethical-moral configuration of our common life is of overriding importance. Here no science can save us. I believe, indeed, that overemphasis on the purely intellectual attitude, often directed solely to the practical and factual, in our education, has led directly to the impairment of ethical values. I am not thinking so much of the dangers with which technical progress has directly confronted mankind, as of the stifling of mutual human considerations by a "matter-of-fact" habit of thought which has come to lie like a killing frost upon human relations.

Fulfillment on the moral and esthetic side is a goal which lies closer to the preoccupations of art than it does to those of science. Of course, *understanding* of our fellow-beings is important. But this understanding becomes fruitful only when it is sustained by sympathetic feeling in joy and in sorrow. The cultivation of this most important spring of moral action is that which is left of religion when it has been purified of the elements of superstition. In this sense, religion forms an important part of education, where it receives far too little consideration, and that little not sufficiently systematic.

The frightful dilemma of the political world situation has much to do with this sin of omission on the part of our civilization. Without "ethical culture" there is no salvation for humanity.

About Education

THE UNIVERSITY COURSES AT DAVOS

In 1928 Einstein participated in the international university courses conducted at Davos, famous Swiss resort for tubercular patients. This address preceded his lecture, "Fundamental Concepts in Physics and Their Development." Published in Mein Weltbild, *Amsterdam: Querido Verlag, 1934.*

Senatores boni viri senatus autem bestia. So a friend of mine, a Swiss professor, once wrote in his facetious way to a university faculty which had annoyed him. Communities tend to be guided less than individuals by conscience and a sense of responsibility. How much misery does this fact cause mankind! It is the source of wars and every kind of oppression, which fill the earth with pain, sighs, and bitterness.

And yet nothing truly valuable can be achieved except by the disinterested cooperation of many individuals. Hence the man of good will is never happier than when some communal enterprise is afoot and is launched at the cost of heavy sacrifices, with the single object of promoting life and culture.

Such pure joy was mine when I heard about the university courses at Davos. A work of rescue is being carried out here, with intelligence and a wise moderation, which is based on a grave need, though it may not be a need that is immediately obvious to everyone. Many a young man goes to this valley with his hopes fixed on the healing power of its sunny mountains and regains his bodily health. But thus withdrawn for long periods from the will-hardening discipline of normal work and a prey to morbid reflection on his physical condition, he easily loses his mental resilience, the sense of being able to hold his own in the struggle for existence. He becomes a sort of hot-house plant and, when his body is cured, often finds it difficult to get back to normal life. This is in particular true for university students. Interruption of intellectual training in the formative period of youth is very apt to leave a gap which can hardly be filled later.

Yet, as a general rule, intellectual work in moderation, so far from retarding cure, indirectly helps it forward, just as moderate physical work will. It is with this realization that the university courses are being instituted for the purpose not merely of preparing these young people for a profession but of stimulating them to intellectual activity as such. They are to provide work, training, and hygiene in the sphere of the mind.

Let us not forget that this enterprise is admirably suited to establish relations between individuals of different nationalities, relations which help to strengthen the idea of a European community. The effects of the new institution in this direction are likely to be all the more advantageous as a result of the fact that the circumstances of its birth rule out every sort of political purpose. The best way to serve the cause of internationalism is by cooperating in some life-giving work.

For all these reasons I rejoice that through the energy and intelligence of the founders, the university courses at Davos have already attained such a measure of success that the enterprise has outgrown the troubles of infancy. May it prosper, enriching the inner lives of numbers of valuable human beings and rescuing many from the poverty of sanatorium life.

TEACHERS AND PUPILS

A talk to a group of children. Published in Mein Weltbild, *Amsterdam: Querido Verlag, 1934.*

MY DEAR CHILDREN:

I rejoice to see you before me today, happy youth of a sunny and fortunate land.

Bear in mind that the wonderful things you learn in your schools are the work of many generations, produced by enthusiastic effort and infinite labor in every country of the world. All this is put into your hands as your inheritance in order that you may receive it, honor it, add to it, and one day faithfully hand it on to your children. Thus do we mortals achieve immortality in the permanent things which we create in common.

If you always keep that in mind you will find a meaning in life and work and acquire the right attitude toward other nations and ages.

EDUCATION AND EDUCATORS

A letter to a young girl. Published in Mein Weltbild, *Amsterdam: Querido Verlag, 1934.*

I have read about sixteen pages of your manuscript and it made me—smile. It is clever, well observed, honest; it stands on its own feet up to a point, and yet it is so typically feminine, by which I mean derivative and steeped in personal resentment. I suffered at the hands of my teachers a similar treatment; they disliked me for my independence and passed me over when they wanted assistants (I must admit, though, that I was somewhat less of a model student than you). But it would not have been worth my while to write anything about my school life, and still less would I have liked to be responsible for anyone's printing or actually reading it. Besides, one always cuts a poor figure if one complains about others who are struggling for their place in the sun, too, after their own fashion.

Therefore, pocket your temperament and keep your manuscript for your sons and daughters, in order that they may derive consolation from it and—not give a damn for what their teachers tell them or think of them.

Incidentally I am only coming to Princeton to do research, not to teach. There is too much education altogether, especially in American schools. The only rational way of educating is to be an example—if one can't help it, a warning example.

EDUCATION AND WORLD PEACE

A message to the Progressive Education Association, November 23, 1934.

The United States, because of its geographic location, is in the fortunate position of being able to teach sane pacifism in the schools, for there exists no serious danger of foreign aggression and hence there is no necessity for inculcating in youth a military spirit. There is, however, a danger that the problem of educating for peace may be handled from an emotional, rather than a realistic standpoint. Little will be gained without a thorough understanding of the underlying difficulties of the problem.

American youth should understand, first of all, that even though actual invasion of American territory is unlikely, the United States is liable to be involved in international entanglements at any time. Reference need only be made to America's participation in the World War to prove the need for such understanding.

Security for the United States, as for other countries, lies only in a satisfactory solution of the world peace problem. Youth must not be allowed to believe that safety can be obtained through political isolation. On the contrary, a serious concern for the *general* peace problem should be aroused. Especially should young people be brought to a clear understanding of how great a responsibility American politicians assumed in failing to support President Wilson's liberal plans at the conclusion of the World War and afterward, thereby hampering the work of the League of Nations in the solution of this problem.

It should be pointed out that nothing can be gained merely by demanding disarmament, so long as there are powerful countries not unwilling to use militaristic methods for the attainment of more advantageous world positions. Moreover, the justification of such proposals as those supported by France, for example, to safeguard individual countries through the establishment of international institutions should be explained. In order to obtain this security, international treaties are needed for common defense against an aggressor. These treaties are necessary, but are not in themselves sufficient. One more step should be taken. Military means of defense should be internationalized, merging and exchanging forces on such a broad scale that military forces stationed in any one country are not withheld for that country's exclusive goals. In preparation for such steps as these, youth must understand the importance of the problem.

The spirit of international solidarity should also be strengthened, and chauvinism should be combated as a hindrance to world peace. In the schools, history should be used as a means of interpreting progress in *civilization,* and not for inculcating ideals of imperialistic power and military success. In my opinion, H. G. Wells' *World History* is to be recommended to students for this point of view. Finally, it is at least of indirect importance that in geography, as well as in history, a sympathetic understanding of the characteristics of various peoples be stimulated, and this understanding should include those peoples commonly designated as "primitive" or "backward."

ON EDUCATION

From an address at Albany, N. Y., on the occasion of the celebration of the tercentenary of higher education in America, October 15, 1936. Translated by Lina Arronet. Published in Out of My Later Years: *New York, Philosophical Library, 1950.*

A day of celebration generally is in the first place dedicated to retrospect, especially to the memory of personages who have gained special distinction for the development of the cultural life. This friendly service for our predecessors must indeed not be neglected, particularly as such a memory of the best of the past is proper to stimulate the well-disposed of today to a courageous effort. But this should be done by someone who, from his youth, has been connected with this State and is familiar with its past, not by one who like a gypsy has wandered about and gathered his experiences in all kinds of countries.

Thus, there is nothing else left for me but to speak about such questions as, independently of space and time, always have been and will be connected with educational matters. In this attempt I cannot lay any claim to being an authority, especially as intelligent and well-meaning men of all times have dealt with educational problems and have certainly repeatedly expressed their views clearly about these matters. From what source shall ı, as a partial layman in the realm of pedagogy, derive courage to expound opinions with no foundations except personal experience and personal conviction? If it were really a scientific matter, one would probably be tempted to silence by such considerations.

However, with the affairs of active human beings it is different. Here knowledge of truth alone does not suffice; on the contrary this knowledge must continually be renewed by ceaseless effort, if it is not to be lost. It resembles a statue of marble which stands in the desert and is continuously threatened with burial by the shifting sand. The hands of service must ever be at work, in order that the marble continue lastingly to shine in the sun. To these serving hands mine also shall belong.

The school has always been the most important means of transferring the wealth of tradition from one generation to the next. This applies today in an even higher degree than in former times, for through modern development of the economic life, the family as bearer of tradition and education has been weakened. The continuance and health of human society is therefore in a still higher degree dependent on the school than formerly.

Sometimes one sees in the school simply the instrument for transferring a certain maximum quantity of knowledge to the growing generation. But that is not right. Knowledge is dead; the school, however, serves the living. It should develop in the young individuals those qualities and capabilities which are of value for the welfare of the commonwealth. But that does not mean that individuality should be destroyed and the individual become a mere tool of the community, like a bee or an ant. For a community of standardized individuals without personal originality and personal aims would be a poor community without possibilities for development. On the contrary, the aim must be the training of independently acting and thinking individuals, who, however, see in the service of the community their highest life problem. So far as I can judge, the English school system comes nearest to the realization of this ideal.

But how shall one try to attain this ideal? Should one perhaps try to realize this aim by moralizing? Not at all. Words are and remain an empty sound, and the road to perdition has ever been accompanied by lip service to an ideal. But personalities are not formed by what is heard and said, but by labor and activity.

The most important method of education accordingly always has consisted of that in which the pupil was urged to actual performance. This applies as well to the first attempts at writing of the primary boy as to the doctor's thesis on graduation from the university, or as to the mere memorizing of a poem, the writing of a composition, the interpretation and translation of a text, the solving of a mathematical problem or the practice of physical sport.

But behind every achievement exists the motivation which is at the foundation of it and which in turn is strengthened and nourished by the accomplishment of the undertaking. Here there are the greatest differences and they are of greatest importance to the educational value of the school. The same work may owe its origin to fear and compulsion, ambitious desire for authority and distinction, or loving interest in the object and a desire for truth and understanding, and thus to that divine curiosity which every healthy child possesses, but which so often is weakened early. The educational influence which is exercised upon the pupil by the accomplishment of one and the same work may be widely different, depending upon whether fear of hurt, egoistic passion, or desire for pleasure and satisfaction is at the bottom of this work. And nobody will maintain that the administration of the school and the attitude of the teachers do not have an influence upon the molding of the psychological foundation for pupils.

To me the worst thing seems to be for a school principally to work with methods of fear, force, and artificial authority. Such treatment destroys the sound sentiments, the sincerity, and the self-confidence of the pupil. It produces the submissive subject. It is no wonder that such schools are the rule in Germany and Russia. I know that the schools in this country are free from this worst evil; this also is so in Switzerland and probably in all democratically governed countries. It is comparatively simple to keep the school free from this worst of all evils. Give into the power of the teacher the fewest possible coercive measures, so that the only source of the pupil's respect for the teacher is the human and intellectual qualities of the latter.

The second-named motive, ambition or, in milder terms, the aiming at recognition and consideration, lies firmly fixed in human nature. With absence of mental stimulus of this kind, human cooperation would be entirely impossible; the desire for the approval of one's fellow-man certainly is one of the most important binding powers of society. In this complex of feelings, constructive and destructive forces lie closely together. Desire for approval and recognition is a healthy motive; but the desire to be acknowledged as better, stronger, or more intelligent than a fellow being or fellow scholar easily leads to an excessively egoistic psychological adjustment, which may become injurious for the individual and for the community. Therefore the school and the teacher must guard against employing the easy method of creating individual ambition, in order to induce the pupils to diligent work.

Darwin's theory of the struggle for existence and the selectivity connected with it has by many people been cited as authorization of the encouragement of the spirit of competition. Some people also in such a way have tried to prove pseudo-scientifically the necessity of the destructive economic struggle of competition between individuals. But this is wrong, because man owes his strength in the struggle for existence to the fact that he is a socially living animal. As little as a battle between single ants of an ant hill is essential for survival, just so little is this the case with the individual members of a human community.

Therefore one should guard against preaching to the young man success in the customary sense as the aim of life. For a successful man is he who receives a great deal from his fellowmen, usually incomparably more than corresponds to his service to them. The value of a man, however, should be seen in what he gives and not in what he is able to receive.

The most important motive for work in the school and in life is the pleasure in work, pleasure in its result, and the knowledge of the value of the result to the community. In the awakening and strengthening of these psychological forces in the young man, I see the most important task given by the school. Such a psychological foundation alone leads to a joyous desire for the highest possessions of men, knowledge and artist-like workmanship.

The awakening of these productive psychological powers is

certainly less easy than the practice of force or the awakening of individual ambition but is the more valuable for it. The point is to develop the childlike inclination for play and the childlike desire for recognition and to guide the child over to important fields for society; it is that education which in the main is founded upon the desire for successful activity and acknowledgment. If the school succeeds in working successfully from such points of view, it will be highly honored by the rising generation and the tasks given by the school will be submitted to as a sort of gift. I have known children who preferred schooltime to vacation.

Such a school demands from the teacher that he be a kind of artist in his province. What can be done that this spirit be gained in the school? For this there is just as little a universal remedy as there is for an individual to remain well. But there are certain necessary conditions which can be met. First, teachers should grow up in such schools. Second, the teacher should be given extensive liberty in the selection of the material to be taught and the methods of teaching employed by him. For it is true also of him that pleasure in the shaping of his work is killed by force and exterior pressure.

If you have followed attentively my meditations up to this point, you will probably wonder about one thing. I have spoken fully about in what spirit, according to my opinion, youth should be instructed. But I have said nothing yet about the choice of subjects for instruction, nor about the method of teaching. Should language predominate or technical education in science?

To this I answer: in my opinion all this is of secondary importance. If a young man has trained his muscles and physical endurance by gymnastics and walking, he will later be fitted for every physical work. This is also analogous to the training of the mind and the exercising of the mental and manual skill. Thus the wit was not wrong who defined education in this way: "Education is that which remains, if one has forgotten everything he learned in school." For this reason I am not at all anxious to take sides in the struggle between the followers of the classical philologic-historical education and the education more devoted to natural science.

On the other hand, I want to oppose the idea that the school has to teach directly that special knowledge and those accomplishments which one has to use later directly in life. The demands of life are much too manifold to let such a specialized training in school appear possible. Apart from that, it seems to me, moreover, objectionable to treat the individual like a dead tool. The school should always have as its aim that the young man leave it as a harmonious personality, not as a specialist. This in my opinion is true in a certain sense even for technical schools, whose students will devote themselves to a quite definite profession. The development of general ability for independent thinking and judgment should always be placed foremost, not the acquisition of special knowledge. If a person masters the fundamentals of his subject and has learned to think and work independently, he will surely find his way and besides will better be able to adapt himself to progress and changes than the person whose training principally consists in the acquiring of detailed knowledge.

Finally, I wish to emphasize once more that what has been said here in a somewhat categorical form does not claim to mean more than the personal opinion of a man, which is founded upon *nothing but* his own personal experience, which he has gathered as a student and as a teacher.

ON CLASSIC LITERATURE

Written for the Jungkaufmann, *a monthly publication of the "Schweizerischer Kaufmaennischer Verein, Jugendbund," February 29, 1952.*

Somebody who reads only newspapers and at best books of contemporary authors looks to me like an extremely near-sighted person who scorns eyeglasses. He is completely dependent on the prejudices and fashions of his times, since he never gets to see or hear anything else. And what a person thinks on his own without being stimulated by the thoughts and experiences of other people is even in the best case rather paltry and monotonous.

There are only a few enlightened people with a lucid mind and style and with good taste within a century. What has been preserved of their work belongs among the most precious possessions of mankind. We owe it to a few writers of antiquity that the people in the Middle Ages could slowly extricate themselves from the superstitions and ignorance that had darkened life for more than half a millennium.

Nothing is more needed to overcome the modernist's snobbishness.

ENSURING THE FUTURE OF MANKIND

Message for Canadian Education Week, March 2–8, 1952.
Published in Mein Weltbild, *Zurich: Europa Verlag, 1953.*

The discovery of nuclear chain reactions need not bring about the destruction of mankind, any more than did the discovery of matches. We only must do everything in our power to safeguard against its abuse. In the present stage of technical development, only a supranational organization, equipped with a sufficiently strong executive power, can protect us. Once we have understood that, we shall find the strength for the sacrifices necessary to ensure the future of mankind. Each one of us would be at fault if the goal were not reached in time. There is the danger that everyone waits idly for others to act in his stead.

The progress of science in our century will be highly appreciated by every knowledgeable person, even by the casual observer who only encounters the technical applications of science. Nevertheless, its recent achievements will not be overrated if the fundamental problems of science are kept in mind. If we ride in a train, we seem to move with incredible speed as long as we watch only nearby objects. But if we direct our attention to prominent features of the landscape, like high mountains, the scenery seems to change very slowly. It is just the same with the fundamental problems in science.

In my opinion, it is not reasonable even to talk of "our way of life" or that of the Russians. In both cases we are dealing with a collection of traditions and customs which do not form an organic whole. It certainly makes more sense to ask which institutions and traditions are harmful, and which are useful, to human beings; which make life happier, or more painful. We then must endeavor to adopt whatever appears best, irrespective of whether, at present, we find it realized at home or somewhere else.

Now to the salaries of teachers. In a healthy society, every useful activity is compensated in a way to permit of a decent living. The exercise of any socially valuable activity gives inner satisfaction; but it cannot be considered as part of the salary. The teacher cannot use his inner satisfaction to fill the stomachs of his children.

EDUCATION FOR INDEPENDENT THOUGHT

From the New York Times, *October 5, 1952.*

It is not enough to teach man a specialty. Through it he may become a kind of useful machine but not a harmoniously developed personality. It is essential that the student acquire an understanding of and a lively feeling for values. He must acquire a vivid sense of the beautiful and of the morally good. Otherwise he—with his specialized knowledge—more closely resembles a well-trained dog than a harmoniously developed person. He must learn to understand the motives of human beings, their illusions, and their sufferings in order to acquire a proper relationship to individual fellow-men and to the community.

These precious things are conveyed to the younger generation through personal contact with those who teach, not—or at least not in the main—through textbooks. It is this that primarily constitutes and preserves culture. This is what I have in mind when I recommend the "humanities" as important, not just dry specialized knowledge in the fields of history and philosophy.

Overemphasis on the competitive system and premature specialization on the ground of immediate usefulness kill the spirit on which all cultural life depends, specialized knowledge included.

It is also vital to a valuable education that independent critical thinking be developed in the young human being, a development that is greatly jeopardized by overburdening him with too much and with too varied subjects (point system). Overburdening necessarily leads to superficiality. Teaching should be such that what is offered is perceived as a valuable gift and not as a hard duty.

About Friends

JOSEPH POPPER-LYNKAEUS

1838–1921. Austrian. By profession, engineer. Famous as a writer for his pungent criticism of State and Society and for his courageous program to alleviate social evils. Some of his books were banned in Imperial Austria. This statement appeared in Mein Weltbild, *Amsterdam: Querido Verlag, 1934.*

Popper-Lynkaeus was more than a brilliant engineer and writer. He was one of the few outstanding personalities who embody the conscience of a generation. He has drummed into us that society is responsible for the fate of every individual and shown us a way to translate the consequent obligation of the community into fact. The community or state was no fetish to him; he based its right to demand sacrifices of the individual entirely on its duty to give the individual personality a chance of harmonious development.

GREETING TO GEORGE BERNARD SHAW

On the occasion of a visit of Einstein's to England in 1930. This message was published in Mein Weltbild, *Amsterdam: Querido Verlag, 1934.*

There are few enough people with sufficient independence to see the weaknesses and follies of their contemporaries and remain themselves untouched by them. And these isolated few usually soon lose their zeal for putting things to rights when they have come face to face with human obduracy. Only to a tiny minority is it given to fascinate their generation by subtle humor and grace and to hold the mirror up to it by the impersonal agency of art. Today I salute with sincere emotion the supreme master of this method, who has delighted—and educated—us all.

IN HONOR OF ARNOLD BERLINER'S
SEVENTIETH BIRTHDAY

From Die Naturwissenschaften, *Vol. 20, p. 913, 1932. Berliner, a German physicist, was editor of that weekly from 1913 to 1935, when, as a Jew, he was deposed by the Nazi regime. Seven years later at the age of eighty, about to be deported, Berliner committed suicide.*

I should like to take this opportunity of telling my friend Berliner and the readers of this periodical why I rate him and his work so highly. It has to be done here because it is our only chance of getting such things said; our training in objectivity has led to a tabu on everything personal, which we mortals may only transgress on quite exceptional occasions such as this.

And now, after this dash of liberty, back to the objectivity! The area of scientific investigation has been enormously extended, and theoretical knowledge has become vastly more profound in every department of science. But the assimilative power of the human intellect is and remains strictly limited. Hence it was inevitable that the activity of the individual investigator should be confined to a smaller and smaller section of human knowledge. Worse still, this specialization makes it increasingly difficult to keep even our general understanding of science as a whole, without which the true spirit of research is inevitably handicapped, in step with scientific progress. A situation is developing similar to the one symbolically represented in the Bible by the story of the tower of Babel. Every serious scientific worker is painfully conscious of this involuntary relegation to an ever-narrowing sphere of knowledge, which threatens to deprive the investigator of his broad horizon and degrades him to the level of a mechanic.

We have all suffered under this evil, without making any effort to mitigate it. But Berliner has come to the rescue, as far as the German-speaking world is concerned, in the most admirable way. He realized that the existing popular periodicals were sufficient to instruct and stimulate the layman; but he also recognized the necessity of a well-balanced periodical directed with particular care for the information of the scientist who desired to familiarize himself with the development in scientific problems, methods, and results in such a way as to be able to form a judgment of his own. Through many years of hard work he has devoted himself to this object with great intelligence and no less great determination, and done us all, and science, a service for which we cannot be too grateful.

It was necessary for him to secure the cooperation of the successful scientists and induce them to say what they had to say in a form as far as possible intelligible to the non-specialist. He often told me of the battles he had to fight in pursuing this objective, describing his difficulties to me in the following riddle: Question: What is a scientific author? Answer: A cross between a mimosa and a porcupine. Berliner's achievement was only possible because his longing for a clear, comprehensive view of as large as possible an area of scientific investigation has remained so strongly alive. This feeling also drove him to produce a textbook of physics, the fruit of many years of strenuous work, of which a medical student said to me the other day: "I don't know how I should ever have got a clear idea of the principles of modern physics in the time at my disposal without this book."

Berliner's fight for clarity and a comprehensive view of science has done a great deal to bring to life in many minds the problems, methods, and results of science. The scientific life of our time is no longer conceivable without his periodical. It is just as important to make knowledge live and to keep it alive as to solve specific problems.

H. A. LORENTZ'S WORK IN THE CAUSE OF INTERNATIONAL COOPERATION

Written in 1927. H. A. Lorentz, a Dutch theoretical physicist, was one of the greatest scientists of his times. His work covered many fields of physics, but his most outstanding contributions were to the theory of electromagnetism in all its ramifications. His discoveries prepared the ground for many of the modern developments in physics, most particularly for the theory of relativity. After World War I, Lorentz put a great deal of effort into the reorganization of international cooperation, particularly among scientists. Owing to his undisputed prestige and the respect which he enjoyed among scholars of all countries, his endeavors met with success. During the last years of his life he was chairman of the League of Nations' Committee of Intellectual Cooperation. This essay appeared in Mein Weltbild, *Amsterdam: Querido Verlag, 1934.*

With the extensive specialization of scientific research which the nineteenth century brought about, it has become rare for a man occupying a leading position in one of the sciences to manage at the same time to do valuable service to the community in the sphere of international organization and international politics. Such service demands not only strength, insight, and a reputation based on solid achievements, but also a freedom from national prejudice and a devotion to the common ends of all, which have become rare in our times. I have met no one who combined all these qualities in himself so perfectly as H. A. Lorentz. The marvelous thing about the effect of his personality was this: Independent and stubborn natures, such as are particularly common among men of learning, do not readily bow to another's will and for the most part only accept his leadership grudgingly. But when Lorentz is in the presidential chair, an atmosphere of happy cooperation is invariably created, however much those present may differ in their aims and habits of thought. The secret of this success lies not only in his swift comprehension of people and things and his marvelous command of language, but above all in this, that one feels that his whole heart is in the business in hand, and that when he is at work, he has room for nothing else in his mind. Nothing disarms the recalcitrant so much as this.

Before the War, Lorentz's activities in the cause of international relations were confined to presiding at congresses of physicists. Particularly noteworthy among these were the Solvay Congresses, the first two of which were held at Brussels in 1909 and 1911. Then came the European war, which was a crushing blow to all who had the improvement of human relations in general at heart. Even before the war was over, and still more after its end, Lorentz devoted himself to the work of international reconciliation. His efforts were especially directed toward the re-establishment of fruitful and friendly cooperation between men of learning and scientific societies. An outsider can hardly conceive what uphill work this was. The accumulated resentment of the war period had not yet died down, and many influential men persisted in the irreconcilable attitude into which they had allowed themselves to be driven by the pressure of circumstances. Lorentz's efforts resembled those of a doctor with a recalcitrant patient who refuses to take the medicines carefully prepared for his benefit.

But Lorentz was not to be deterred, once he had recognized a course of action as the right one. Right after the war, he joined the governing body of the "Conseil de recherche" which was founded by the scholars of the victorious countries, and from which the scholars and learned societies of the Central Powers were excluded. His object in taking this step, which caused great offense to the academic world of the Central Powers, was to influence this institution in such a way that it could be expanded into something truly international. He and other right-minded men succeeded, after repeated efforts, in securing the removal of the offensive exclusion-clause from the statutes of the "Conseil." The goal, which was the restoration of normal and fruitful cooperation between learned societies, is, however, not yet attained, because the academic world of the Central Powers, exasperated by nearly ten years of exclusion from practically all international scientific gatherings, has got into a habit of keeping itself to itself. Now, however, there are good grounds for hoping that the ice will soon be broken, thanks to the tactful efforts of Lorentz, prompted by pure enthusiasm for the good cause.

Lorentz has also devoted his energies to the service of international cultural ends in another way, by consenting to serve on the League of Nations' Committee of Intellectual Cooperation, which was called into existence some five years ago with Bergson as chairman. For the last year Lorentz has presided over the Committee, which, with the active support of its subordinate, the Paris Institute, is to act as a go-between in the domain of intellectual and artistic activity among the various spheres of culture. There, too, the beneficent influence of this wise, humane, and modest personality, whose unspoken but faithfully followed device is, "Not mastery but service," will lead people on the right way.

May his example contribute to the triumph of that spirit!

ADDRESS AT THE GRAVE OF H. A. LORENTZ

Lorentz, born 1853, died 1928. This address was published in Mein Weltbild, *Amsterdam: Querido Verlag, 1934.*

It is as the representative of the German-speaking academic world and in particular the Prussian Academy of Sciences, but above all as a pupil and affectionate admirer that I stand at the grave of the greatest and noblest man of our times. His genius led the way from Maxwell's work to the achievements of contemporary physics, to which he contributed important building stones and methods.

He shaped his life like an exquisite work of art down to the smallest detail. His never-failing kindness and generosity and his sense of justice, coupled with a sure and intuitive understanding of people and human affairs, made him a leader in any sphere he entered. Everyone followed him gladly, for they felt that he never set out to dominate but only to serve. His work and his example will live on as an inspiration and a blessing to many generations.

H. A. LORENTZ, CREATOR AND PERSONALITY

Message delivered at Leyden, Holland, 1953, for the commemoration of the one hundredth anniversary of the birth of Lorentz. Published in Mein Weltbild, *Zurich: Europa Verlag, 1953.*

At the turn of the century the theoretical physicists of all nations considered H. A. Lorentz as the leading mind among them, and rightly so. The physicists of our time are mostly not fully aware of the decisive part which H. A. Lorentz played in shaping the fundamental ideas in theoretical physics. The reason for this strange fact is that Lorentz's basic ideas have become so much a part of them that they are hardly able to realize quite how daring these ideas have been and to what extent they have simplified the foundations of physics.

When H. A. Lorentz started his creative scientific work, Maxwell's theory of electromagnetism had already won out. But there was inherent in this theory a peculiar complexity of the fundamental principles which prevented its essential features from revealing themselves distinctly. Though the field concept had indeed displaced the concept of action at a distance, the electric and magnetic fields were not yet conceived as primary entities, but rather as states of ponderable matter which latter was treated as a continuum. Consequently the electric field appeared decomposed into the electric field strength and the dielectric displacement. In the simplest case, these two fields were connected by the dielectric constant, but in principle they were considered and treated as independent entities. The magnetic field was treated similarly. It was in accordance with this basic idea to treat empty space as a special case of ponderable matter in which the relation between field strength and displacement happened to be particularly simple. In particular, this interpretation brought it about that the electric and magnetic field could not be conceived independent of the state of motion of matter, which was considered the carrier of the field.

A good idea of the interpretation of Maxwell's electrodynamics then prevailing may be gained from the study of H. Hertz's investigation on the electrodynamics of moving bodies.

Then came H. A. Lorentz's decisive simplification of the theory. He based his investigations with unfaltering consistency upon the following hypotheses:

The seat of the electromagnetic field is the empty space. In it there are only *one* electric and *one* magnetic field vector. This field is generated by atomistic electric charges upon which the field in turn exerts ponderomotive forces. The only connection between the electromagnetic field and ponderable matter arises from the fact that elementary electric charges are rigidly attached to atomistic particles of matter. For the latter Newton's law of motion holds.

Upon this simplified foundation Lorentz based a complete

theory of all electromagnetic phenomena known at the time, including those of the electrodynamics of moving bodies. It is a work of such consistency, lucidity, and beauty as has only rarely been attained in an empirical science. The only phenomenon that could not be entirely explained on this basis, i.e., without additional assumptions, was the famous Michelson-Morley experiment. Without the localization of the electromagnetic field in empty space this experiment could not conceivably have led to the theory of special relativity. Indeed, the essential step was just the reduction of electromagnetism to Maxwell's equations in empty space or—as it was expressed at that time—in ether.

H. A. Lorentz even discovered the "Lorentz transformation," later called after him, though without recognizing its group character. To him Maxwell's equations in empty space held only for a particular coordinate system distinguished from all other coordinate systems by its state of rest. This was a truly paradoxical situation because the theory seemed to restrict the inertial system more strongly than did classical mechanics. This circumstance, which from the empirical point of view appeared completely unmotivated, was bound to lead to the theory of special relativity.

Thanks to the generosity of the University of Leiden, I frequently spent some time there staying with my dear and unforgettable friend, Paul Ehrenfest. Thus I had often the opportunity to attend Lorentz's lectures which he gave regularly to a small circle of young colleagues after he had already retired from his professorship. Whatever came from this supreme mind was as lucid and beautiful as a good work of art and was presented with such facility and ease as I have never experienced in anybody else.

If we younger people had known H. A. Lorentz only as a sublime mind, our admiration and respect for him would have been unique. But what I feel when I think of H. A. Lorentz is far more than that. He meant more to me personally than anybody else I have met in my lifetime.

Just as he was in command of physics and of the mathematical formalism, thus he also was in command of himself without effort and strain. His quite unusual lack of human frailties never had a depressing effect on others. Everybody felt his superiority, but nobody felt oppressed by it. Though he had no illusions about people and human affairs, he was full of kindness toward everybody and everything. Never did he give the impression of domineering, always of serving and helping. He was extremely conscientious without allowing anything to assume undue importance; a subtle humor guarded him, which was reflected in his eyes and in his smile. And it fits that, not-withstanding all his devotion to science, he was convinced that our comprehension cannot penetrate too deeply into the essence of things. Only in my later years was I able to appreciate fully this half-skeptical, half-humble attitude.

In spite of my honest attempts I find that language—or at least my language—cannot do justice to the subject of this short piece of writing. Therefore I shall only quote two short sayings of Lorentz's that impressed me particularly deeply:

"I am happy to belong to a nation that is too small to commit big follies."

To a man who in a conversation during the first World War tried to convince him that in the human sphere fate is determined by might and force he gave this reply:

"It is conceivable that you are right. But I would not want to live in such a world."

MARIE CURIE IN MEMORIAM

Statement for the Curie Memorial Celebration, Roerich Museum, New York, November 23, 1935. Published in Out of My Later Years, *New York: Philosophical Library, 1950.*

At a time when a towering personality like Mme. Curie has come to the end of her life, let us not merely rest content with recalling what she has given to mankind in the fruits of her work. It is the moral qualities of its leading personalities that are perhaps of even greater significance for a generation and for the course of history than purely intellectual accomplishments. Even these latter are, to a far greater degree than is commonly credited, dependent on the stature of character.

It was my good fortune to be linked with Mme. Curie through twenty years of sublime and unclouded friendship. I came to admire her human grandeur to an ever growing degree. Her strength, her purity of will, her austerity toward herself, her objectivity, her incorruptible judgment—all these were of a kind seldom found joined in a single individual. She felt herself at every moment to be a servant of society, and her profound modesty never left any room for complacency. She was oppressed by an abiding sense for the asperities and inequities of society. This is what gave her that severe outward aspect, so easily misinterpreted by those who were not close to her—a curious severity unrelieved by any artistic strain. Once she had recognized a certain way as the right one, she pursued it without compromise and with extreme tenacity.

The greatest scientific deed of her life—proving the existence of radioactive elements and isolating them—owes its accomplishment not merely to bold intuition but to a devotion and tenacity in execution under the most extreme hardships imaginable, such as the history of experimental science has not often witnessed.

If but a small part of Mme. Curie's strength of character and devotion were alive in Europe's intellectuals, Europe would face a brighter future.

MAHATMA GANDHI

On the occasion of Gandhi's seventieth birthday in 1939. Published in Out of My Later Years, *New York: Philosophical Library, 1950.*

A leader of his people, unsupported by any outward authority: a politician whose success rests not upon craft nor the

mastery of technical devices, but simply on the convincing
power of his personality; a victorious fighter who has always
scorned the use of force; a man of wisdom and humility, armed
with resolve and inflexible consistency, who has devoted all his
strength to the uplifting of his people and the betterment of
their lot; a man who has confronted the brutality of Europe
with the dignity of the simple human being, and thus at all
times risen superior.

Generations to come, it may be, will scarce believe that such
a one as this ever in flesh and blood walked upon this earth.

MAX PLANCK IN MEMORIAM

*Read at the Max Planck Memorial Services, 1948. Pub-
lished in* Out of My Later Years, *New York: Philosophical
Library, 1950.*

A man to whom it has been given to bless the world with
a great creative idea has no need for the praise of posterity. His
very achievement has already conferred a higher boon upon
him.

Yet it is good—indeed, it is indispensable—that representatives of all who strive for truth and knowledge should be
gathered here today from the four corners of the globe. They
are here to bear witness that even in these times of ours, when
political passion and brute force hang like swords over the
anguished and fearful heads of men, the standard of our ideal
search for truth is being held aloft undimmed. This ideal, a
bond forever uniting scientists of all times and in all places, was
embodied with rare completeness in Max Planck.

Even the Greeks had already conceived the atomistic nature
of matter and the concept was raised to a high degree of probability by the scientists of the nineteenth century. But it was
Planck's law of radiation that yielded the first exact determination—independent of other assumptions—of the absolute magnitudes of atoms. More than that, he showed convincingly that

in addition to the atomistic structure of matter there is a kind of atomistic structure to energy, governed by the universal constant h, which was introduced by Planck.

This discovery became the basis of all twentieth-century research in physics and has almost entirely conditioned its development ever since. Without this discovery it would not have been possible to establish a workable theory of molecules and atoms and the energy processes that govern their transformations. Moreover, it has shattered the whole framework of classical mechanics and electrodynamics and set science a fresh task: that of finding a new conceptual basis for all physics. Despite remarkable partial gains, the problem is still far from a satisfactory solution.

In paying homage to this man, the American National Academy of Sciences expresses its hope that free research, for the sake of pure knowledge, may remain unhampered and unimpaired.

MESSAGE IN HONOR OF
MORRIS RAPHAEL COHEN

For the Morris Raphael Cohen Student Memorial Fund, November 15, 1949.

LADIES AND GENTLEMEN:

It was a pleasure to learn that there are people in the turbulent metropolis who are not completely absorbed by the obtrusive impressions of the moment. Your symposium bears witness that the relations among thinking human beings are threatened neither by the pretentious present nor by the dividing line of death. The majority of those who are particularly close to us are no longer among the living; Morris Cohen has been lately included in their number.

I knew him well as an extraordinarily helpful, conscientious man of unusually independent character and I rather frequently had the pleasure of discussing with him problems of common interest. But when I occasionally tried to tell something about his spiritual personality, I realized painfully that I was not acquainted enough with the working of his mind.

To fill this lacuna—at least scantily—I took his book *Logic and Scientific Method*, which he had published jointly with Ernest Nagel. I did not do this comfortably but with a well-founded unrest because there was so little time. But when I had started reading, I became so fascinated that the external occasion of my reading receded somewhat into the background.

When, after several hours, I came to myself again, I asked myself what it was that had so fascinated me. The answer is simple. The results were not presented as ready-made, but scientific curiosity was first aroused by presenting contrasting possibilities of conceiving the matter. Only then the attempt was made to clarify the issue by thorough argument. The intellectual honesty of the author makes us share the inner struggle in his mind. It is this which is the mark of the born teacher. Knowledge exists in two forms—lifeless, stored in books, and alive in the consciousness of men. The second form of existence is after all the essential one; the first, indispensable as it may be, occupies only an inferior position.

PART II

ON POLITICS, GOVERNMENT, AND PACIFISM

THE INTERNATIONAL OF SCIENCE

Written shortly after World War I. Published in Mein Weltbild, *Amsterdam: Querido Verlag, 1934.*

At a sitting of the Academy during the War, at the time when nationalism and political infatuation had reached its height, Emil Fischer spoke the following emphatic words: "It's no use, gentlemen, science is and remains international." The really great scientists have always known this and felt it passionately, even though in times of political strife they may have remained isolated among their colleagues of inferior caliber. In every camp during the War this group of voters betrayed their sacred trust. The International Association of Academies was broken up. Congresses were and still are held from which colleagues from ex-enemy countries are excluded. Political considerations, advanced with much solemnity, prevent the triumph of the purely objective ways of thinking without which our great aims must necessarily be frustrated.

What can right-minded people, people who are proof against the emotional temptations of the moment, do to repair the damage? With the majority of intellectual workers still so excited, truly international congresses on the grand scale cannot yet be held. The psychological obstacles to the restoration of the international associations of scientific workers are still too formidable to be overcome by the minority whose ideas and feelings are of a more comprehensive kind. Men of this kind can aid in the great work of restoring the international societies to health by keeping in close touch with like-minded people all over the world, and steadfastly championing the international cause in their own spheres. Success on a large scale will take time, but it will undoubtedly come. I cannot let this opportunity pass without paying tribute, in particular, to the large

83

number of our English colleagues whose desire to preserve the confraternity of the intellect has remained alive through all these difficult years.

The attitude of the individual is everywhere better than the official pronouncements. Right-minded people should bear this in mind and not allow themselves to be exasperated or misled: *senatores boni viri, senatus autem bestia.*

If I am full of confident hope concerning the progress of international organization, that feeling is based not so much on my confidence in the intelligence and high-mindedness of my fellows, but rather on the imperative pressure of economic developments. And since these depend largely on the work even of reactionary scientists, they, too, will help to create the international organization against their wills.

A FAREWELL

A letter written in 1923 regarding Einstein's resignation from the League of Nations' Committee of Intellectual Co-operation, in protest at the inadequacy of the League. Albert Dufour-Feronce, at the time a high official in the German Foreign Office, later became first German Under-Secretary of the League of Nations. In 1924 Einstein, to counteract the exploitation of his earlier decision by German chauvinists in their propaganda against international cooperation, rejoined the Committee of Intellectual Co-operation. Published in Mein Weltbild, *Amsterdam: Querido Verlag, 1934.*

DEAR MR. DUFOUR-FERONCE:

Your kind letter must not go unanswered, otherwise you may get a mistaken notion of my attitude. The grounds for my resolve to go to Geneva no more are as follows: experience has, unhappily, taught me that the Commission, taken as a whole, stands for no serious determination to make real progress in the task of improving international relations. It looks to me far more like an embodiment of the principle *ut aliquid fieri videatur.* The Commission seems to me even worse in this respect than the League taken as a whole.

It is precisely because I desire to work with all my might for the establishment of an international arbitrating and regulative authority *superior to the state,* and because I have this object so very much at heart, that I feel compelled to leave the Commission.

The Commission has given its blessing to the oppression of the cultural minorities in all countries by causing a National Commission to be set up in each of them, which is to form the only channel of communication between the intellectuals of a country and the Commission. It has thereby deliberately abandoned its function of giving moral support to the national minorities in their struggle against cultural oppression.

Further, the attitude of the Commission in the matter of combating the chauvinistic and militaristic tendencies of education in the various countries has been so lukewarm that no serious efforts in this fundamentally important sphere can be hoped for from it.

The Commission has invariably failed to give moral support to those individuals and associations who have thrown themselves without reserve into the task of working for an international order and against the military system.

The Commission has never made any attempt to resist the appointment of members whom it knew to stand for tendencies the very reverse of those they were bound in duty to advance.

I will not bother you with any further arguments, since you will understand my resolve well enough from these few hints. It is not my business to draw up an indictment but merely to explain my position. If I nourished any hope whatever I should act differently—of that you may be sure.

THE INSTITUTE OF INTELLECTUAL
COOPERATION

Probably written in 1926. Published in Mein Weltbild,
Amsterdam: Querido Verlag, 1934.

During this year the leading politicians of Europe have for
the first time drawn the logical conclusion from the realization
that our continent can only regain its prosperity if the latent
struggle between the traditional political units ceases. The
political organization of Europe must be strengthened, and a
gradual attempt made to abolish tariff barriers. This great end
cannot be achieved by treaties alone. The minds of the people
must, above all, be prepared for it. We must try gradually to
awaken in them a sense of solidarity which will not, as heretofore, stop at frontiers. It is with this in mind that the League of
Nations created the *Commission de co-opération intellec-
tuelle.* This commission was to be a strictly international and
entirely non-political body, whose business it was to put the intellectuals of all the nations, who were isolated by the War, in
touch with each other. It proved a difficult task; for it has, alas,
to be admitted that—at least in the countries with which I am
most closely acquainted—the artists and men of learning permit
themselves to be governed by narrow nationalism to a far greater
extent than the men of affairs.

Hitherto this commission has met twice a year. To make its
efforts more effective, the French government has decided to
create and maintain a permanent Institute of Intellectual Cooperation, which is just now to be opened. It is a generous act
on the part of the French government and as such deserves the
thanks of all.

It is an easy and grateful task to rejoice and praise and to say
nothing about the things one regrets or disapproves of. But
honesty alone can help our work forward, so I will not shrink
from combining criticism with this greeting to the newborn
child.

I have daily occasion for observing that the greatest obstacle which the work of our commission has to encounter is the lack of confidence in its political impartiality. Everything must be done to strengthen that confidence and anything avoided that might harm it.

When, therefore, the French government sets up and maintains an Institute out of public funds in Paris as a permanent organ of the Commission, with a Frenchman as its Director, the outside observer can hardly avoid the impression that French influence predominates in the Commission. This impression is further strengthened by the fact that a Frenchman has also been chairman of the Commission itself thus far. Although the individuals in question are men of the highest reputation, esteemed and respected everywhere, nevertheless the impression remains.

Dixi et salvavi animam meam. I hope with all my heart that the new Institute by constant interaction with the Commission will succeed in promoting their common ends and winning the confidence and recognition of intellectual workers all over the world.

THOUGHTS ON THE WORLD ECONOMIC CRISIS

This and the following two articles were written during the world economic crisis of the 1930's. Although prevailing conditions are not the same and some of the suggested remedies have been used by various countries, these articles should be included. Published in Mein Weltbild, *Amsterdam: Querido Verlag, 1934.*

If there is anything that can give a layman in the sphere of economics the courage to express an opinion on the nature of the alarming economic difficulties of the present day, it is the hopeless confusion of opinions among the experts. What I have to say is nothing new and does not pretend to be anything more than the expression of the opinion of an independent and honest man who, unburdened by class or national prejudices, desires nothing but the good of humanity and the most harmonious possible scheme of human existence. If in what follows I write as if I were sure of the truth of what I am saying, this is merely done for the sake of an easier mode of expression; it does not proceed from unwarranted self-confidence or a belief in the infallibility of my somewhat simple intellectual conception of problems which are in reality uncommonly complex.

As I see it, this crisis differs in character from past crises in that it is based on an entirely new set of conditions, arising out of the rapid progress in methods of production. Only a fraction of the available human labor in the world is now needed for the production of the total amount of consumption goods necessary to life. Under a completely laissez-faire economic system, this fact is bound to lead to unemployment.

For reasons which I do not propose to analyze here, the majority of people are compelled to work for the minimum wage on which life can be supported. If two factories produce the same sort of goods, other things being equal, that factory will be able to produce them more cheaply which employs fewer workmen—i.e., makes the individual worker work as long and as hard as human nature permits. From this it follows inevitably that, with methods of production as they are today, only a portion of the available labor can be used. While unreasonable demands are made on this portion, the remainder is automatically excluded from the process of production. This leads to a fall in sales and profits. Businesses go smash, which further increases unemployment and diminishes confidence in industrial concerns and therewith public participation in the mediating banks; finally the banks become insolvent through the sudden withdrawal of accounts and the wheels of industry therewith come to a complete standstill.

The crisis has also been attributed to other causes which we will now consider.

Over-production. We have to distinguish between two things here—real over-production and apparent over-production. By real over-production I mean a production so great that it exceeds the demand. This may perhaps apply to motor cars and wheat in the United States at the present moment, although even that is doubtful. By "over-production" people usually mean a condition in which more of one particular article is produced than can, in existing circumstances, be sold, in spite of a shortage of consumption goods among consumers. This I call apparent over-production. In this case it is not the demand that is lacking but the consumers' purchasing-power. Such apparent over-production is only another word for a crisis and therefore cannot serve as an explanation of the latter; hence people who try to make over-production responsible for the present crisis are merely juggling with words.

Reparations. The obligation to pay reparations lies heavy on the debtor nations and their economies. It compels them to go in for dumping and so harms the creditor-nations too. This is beyond dispute. But the appearance of the crisis in the United States, in spite of the high tariff-wall, proves that this cannot be the principal cause of the world crisis. The shortage of gold in the debtor countries due to reparations can at most serve as an argument for putting an end to these payments; it cannot provide an explanation of the world crisis.

Erection of new tariff-walls. Increase in the unproductive burden of armaments. Political insecurity owing to latent danger of war. All these things make the situation in Europe considerably worse without really affecting America. The appearance of the crisis in America shows that they cannot be its principal causes.

The dropping-out of the two powers, China and Russia. Also this blow to world trade cannot make itself very deeply felt in America and therefore cannot be the principal cause of the crisis.

The economic rise of the lower classes since the War. This, supposing it to be a reality, could only produce a scarcity of goods, not an excessive supply.

I will not weary the reader by enumerating further contentions which do not seem to me to get to the heart of the matter. Of one thing I feel certain: this same technical progress which, in itself, might relieve mankind of a great part of the labor necessary to its subsistence, is the main cause of our present misery. Hence there are those who would in all seriousness forbid the introduction of technical improvements. This is obviously absurd. But how can we find a more rational way out of our dilemma?

If we could somehow manage to prevent the purchasing-power of the masses, measured in terms of goods, from sinking below a certain minimum, stoppages in the industrial cycle such as we are experiencing today would be rendered impossible.

The logically simplest but also most daring method of achieving this is a completely planned economy, in which consumption goods are produced and distributed by the community. That is essentially what is being attempted in Russia today. Much will depend on what results this forced experiment produces. To hazard a prophecy here would be presumption. Can goods be produced as economically under such a system as under one which leaves more freedom to individual enterprise? Can this system maintain itself at all without the terror that has so far accompanied it, to which none of us westerners would care to expose himself? Does not such a rigid, centralized economic system tend toward protectionism and toward resistance to advantageous innovations? We must take care, however, not to allow these misgivings to become prejudices which prevent us from forming an objective judgment.

My personal opinion is that those methods are in general preferable which respect existing traditions and habits so far as that is in any way compatible with the end in view. Nor do I believe that a sudden transference of economy into governmental management would be beneficial from the point of view of production; private enterprise should be left its sphere of activity, in so far as it has not already been eliminated by industry itself by the device of cartelization.

There are, however, two respects in which this economic freedom ought to be limited. In each branch of industry the number of working hours per week ought so to be reduced by law that unemployment is systematically abolished. At the same time minimum wages must be fixed in such a way that the purchasing power of the workers keeps pace with production.

Further, in those industries which have become monopolistic in character through organization on the part of the producers, prices must be controlled by the state in order to keep the issue of capital within reasonable bounds and prevent the artificial strangling of production and consumption.

In this way it might perhaps be possible to establish a proper balance between production and consumption without too great a limitation of free enterprise and at the same time to stop the intolerable tyranny of the owners of the means of production (land and machinery) over the wage-earners, in the widest sense of the term.

PRODUCTION AND PURCHASING POWER

Mein Weltbild, *Amsterdam: Querido Verlag, 1934.*

I do not believe that the remedy for our present difficulties lies in a knowledge of productive capacity and consumption, because this knowledge is likely, in the main, to come too late. Moreover, the trouble in Germany seems to me to be not hypertrophy of the machinery of production but deficient purchasing power in a large section of the population, which has been cast out of the productive process through the rationalization of industry.

The gold standard has, in my opinion, the serious disadvantage that a shortage in the supply of gold automatically leads to a contraction of credit and also of the amount of currency in circulation, to which contraction prices and wages cannot adjust themselves sufficiently quickly.

The natural remedies of our troubles are, in my opinion, as follows:

(1) A statutory reduction of working hours, graduated for each department of industry, in order to get rid of unemployment, combined with the fixing of minimum wages for the purpose of adjusting the purchasing-power of the masses to the amount of goods available.

(2) Control of the amount of money in circulation and of the volume of credit in such a way as to keep the price level steady, abolishing any monetary standard.

(3) Statutory limitation of prices for such articles as have been practically withdrawn from free competition by monopolies or the formation of cartels.

PRODUCTION AND WORK

Answer to a communication. Published in Mein Weltbild, *Amsterdam: Querido Verlag, 1934.*

The fundamental trouble seems to me to be the almost unlimited freedom of the labor market combined with extraordinary progress in the methods of production. To satisfy the needs of the world today nothing like all the available labor is wanted. The result is unemployment and unhealthy competition among the workers, both of which reduce purchasing-power and thereby put the whole economic system intolerably out of gear.

I know Liberal economists maintain that every economy in labor is counterbalanced by an increase in demand. But, to begin with, I don't believe that; and even if it were true, the above-mentioned factors would always operate to force the standard of living of a large portion of the human race down to an unnaturally low level.

I also share your conviction that steps absolutely must be taken to make it possible and necessary for the younger people to take part in the productive process. Further, that the older people ought to be excluded from certain sorts of work (which I call "unqualified" work), receiving instead a certain income, as having by that time done enough work of a kind accepted by society as productive.

I, too, am in favor of abolishing large cities, but not of settling people of a particular type, e.g., old people, in particular towns. Frankly, the idea strikes me as horrible.

I am also of the opinion that fluctuations in the value of money must be avoided, by substituting for the gold standard a standard based on certain classes of goods selected according to the conditions of consumption—as Keynes, if I am not mistaken, long ago proposed. With the introduction of this system one might consent to a certain amount of "inflation," as compared with the present monetary situation, if one could believe that the state would really make a rational use of the windfall thus accruing to it.

The weaknesses of your plan lie, so it seems to me, in the sphere of psychology, or rather, in your neglect of it. It is no accident that capitalism has brought with it progress not merely in production but also in knowledge. Egoism and competition are, alas, stronger forces than public spirit and sense of duty. In Russia, they say, it is impossible to get a decent piece of bread. . . . Perhaps I am over-pessimistic concerning state and other forms of communal enterprise, but I expect little good from them. Bureaucracy is the death of any achievement. I have seen and experienced too many dreadful warnings, even in comparatively model Switzerland.

I am inclined to the view that the state can only be of real use to industry as a limiting and regulative force. It must see to it that competition among the workers is kept within healthy limits, that all children are given a chance to develop soundly, and that wages are high enough for the goods produced to be consumed. But it can exert a decisive influence through its regulative function if its measures are framed in an objective spirit by independent experts.

ADDRESS TO THE STUDENTS' DISARMAMENT MEETING

Delivered before a group of German pacifist students, about 1930. Published in Mein Weltbild, *Amsterdam: Querido Verlag, 1934.*

Preceding generations have presented us with a highly developed science and technology, a most valuable gift which carries with it possibilities of making our life free and beautiful to an extent such as no previous generation has enjoyed. But this gift also brings with it dangers to our existence as great as any that have ever threatened it.

The destiny of civilized humanity depends more than ever on the moral forces it is capable of generating. Hence the task that confronts our age is certainly no easier than the tasks our immediate predecessors successfully performed.

The necessary supply of food and consumer goods can be produced in far fewer hours of work than formerly. Moreover, the problem of distribution of labor and of manufactured goods has become far more difficult. We all feel that the free play of economic forces, the unregulated and unrestrained pursuit of wealth and power by the individual, no longer leads automatically to a tolerable solution of these problems. Production, labor, and distribution need to be organized on a definite plan, in order to prevent the elimination of valuable productive energies and the impoverishment and demoralization of large sections of the population.

If unrestricted sacred egoism leads to dire consequences in economic life, it is still worse as a guide in international relations. The development of mechanical methods of warfare is such that human life will become intolerable if people do not discover before long a way of preventing war. The importance of this object is only equaled by the inadequacy of the attempts hitherto made to attain it.

People seek to minimize the danger by limitation of armaments and restrictive rules for the conduct of war. But war is not a parlor game in which the players obediently stick to the rules. Where life and death are at stake, rules and obligations go by the board. Only the absolute repudiation of all war can be of any use here. The creation of an international court of arbitration is not enough. There must be treaties guaranteeing that the decisions of this court shall be made effective by all the nations acting in concert. Without such a guarantee the nations will never have the courage to disarm seriously.

Suppose, for example, that the American, English, German, and French governments insisted that the Japanese government put an immediate stop to their warlike operations in China, under pain of a complete economic boycott. Do you suppose that any Japanese government would be found ready to take the responsibility of plunging its country into the perilous adventure of defying this order? Then why is it not done? Why must every individual and every nation tremble for their existence? Because each seeks his own wretched momentary advantage and refuses to subordinate it to the welfare and prosperity of the community.

That is why I began by telling you that the fate of the human race was more than ever dependent on its moral strength today. The way to a joyful and happy existence is everywhere through renunciation and self-limitation.

Where can the strength for such a process come from? Only from those who have had the chance in their early years to fortify their minds and broaden their outlook through study. Thus we of the older generation look to you and hope that you will strive with all your might and achieve what was denied to us.

THE DISARMAMENT CONFERENCE OF 1932

From The Nation, *Vol. 133, p. 300. 1931. Original German text published in* Mein Weltbild, *Amsterdam: Querido Verlag, 1934.*

I.

May I begin with an article of political faith? It runs as follows: the state is made for man, not man for the state. The same may be said of science. These are old sayings, coined by men for whom human personality has the highest human value. I should shrink from repeating them, were it not that they are forever threatening to fall into oblivion, particularly in these days of organization and stereotypes. I regard it as the chief duty of the state to protect the individual and give him the opportunity to develop into a creative personality.

That is to say, the state should be our servant and not we its slaves. The state transgresses this commandment when it compels us by force to engage in military and war service, the more so since the object and the effect of this slavish service is to kill people belonging to other countries or interfere with their freedom of development. We are only to make such sacrifices to the state as will promote the free development of individual human beings. To every American all this may be a platitude, but not to every European. Hence we may hope that the fight against war will find strong support among Americans.

And now for the Disarmament Conference. Ought one to laugh, weep, or hope when one thinks of it? Imagine a city inhabited by fiery-tempered, dishonest, and quarrelsome citizens. The constant danger to life there is felt as a serious handicap which makes all healthy development impossible. The City Council desires to remedy this abominable state of affairs, although all the counselors and the rest of the citizens insist on continuing to carry a dagger in their belts. After years of preparation the City Council determines to compromise and raises the question, how long and how sharp the dagger is allowed to be which anyone may carry in his belt when he goes for a walk. As long as the cunning citizens do not suppress knifing by legislation, the courts, and the police, things go on in the old way, of course. A definition of the length and sharpness of the permitted dagger will only help the strongest and most turbulent and leave the weaker at their mercy. You will all understand the meaning of this parable. It is true that we have a League of Nations and a Court of Arbitration. But the League is not much more than a meeting-place and the Court has no means of enforcing its decisions. These institutions provide no security for any country in case of an attack upon it. If you bear this in mind, you will judge the attitude of the French, their refusal to disarm without security, less harshly than it is usually judged at present.

Unless we can agree to limit the sovereignty of the individual state by binding every one of them to take joint action against any country which openly or secretly resists a judgment of the

Court of Arbitration, we shall never get out of a state of universal anarchy and terror. No sleight of hand can reconcile the unlimited sovereignty of the individual country with security against attack. Will it need new disasters to induce the countries to undertake to enforce every decision of the recognized international court? The progress of events so far scarcely justifies us in hoping for anything better in the near future. But everyone who cares for civilization and justice must exert all his strength to convince his fellows of the necessity for laying all countries under an international obligation of this kind.

It will be urged against this notion, not without a certain justification, that it overestimates the efficacy of machinery, and neglects the psychological, or rather the moral, factor. Spiritual disarmament, people insist, must precede material disarmament. They say further, and truly, that the greatest obstacle to international order is that monstrously exaggerated spirit of nationalism which also goes by the fair-sounding but misused name of patriotism. During the last century and a half this idol has acquired an uncanny and exceedingly pernicious power everywhere.

To estimate this objection at its proper worth, one must realize that a reciprocal relation exists between external machinery and internal states of mind. Not only does the machinery depend on traditional modes of feeling and owe its origin and its survival to them, but the existing machinery in its turn exercises a powerful influence on national modes of feeling.

The present deplorably high development of nationalism everywhere is, in my opinion, intimately connected with the institution of compulsory military service or, to call it by its sweeter name, national armies. A state which demands military service of its inhabitants is compelled to cultivate in them a nationalistic spirit, thereby laying the psychological foundation for their military usefulness. In its schools it must idolize, alongside with religion, its instrument of brutal force in the eyes of the youth.

The introduction of compulsory military service is therefore, to my mind, the prime cause of the moral decay of the white race, which seriously threatens not merely the survival of our civilization but our very existence. This curse, along with great social blessings, started with the French Revolution, and before long dragged all the other nations in its train.

Therefore, those who desire to cultivate an international spirit and to combat chauvinism must take their stand against compulsory military service. Is the severe persecution to which conscientious objectors to military service are subjected today a whit less disgraceful to the community than those to which the martyrs of religion were exposed in former centuries? Can you, as the Kellogg Pact does, condemn war and at the same time leave the individual to the tender mercies of the war machine in each country?

If, in view of the Disarmament Conference, we are not merely to restrict ourselves to the technical problems of organization, but also to tackle the psychological question more directly from the standpoint of educational motives, we must try along international lines to create legal means by which the individual can refuse to serve in the army. Such a regulation would undoubtedly produce a great moral effect.

Let me summarize my views: Mere agreements to limit armaments furnish no sort of security. Compulsory arbitration must be supported by an executive force, guaranteed by all the participating countries, which is ready to proceed against the disturber of the peace with economic and military sanctions. Compulsory military service, as the hotbed of unhealthy nationalism, must be combated; most important of all, conscientious objectors must be protected on an international basis.

II.

The benefits that the inventive genius of man has conferred on us in the last hundred years could make life happy and carefree, if organization had been able to keep pace with technical progress. As it is, in the hands of our generation these hard-won achievements are like a razor wielded by a child of three. The possession of marvelous means of production has brought care and hunger instead of freedom.

The results of technical progress are most baleful where they furnish means for the destruction of human life and the hard-won fruits of toil, as we of the older generation experienced to our horror in the World War. More dreadful even than the destruction, in my opinion, is the humiliating slavery into which war plunges the individual. Is it not a terrible thing to be forced by society to do things which all of us as individuals regard as abominable crimes? Only a few had the moral greatness to resist; them I regard as the real heroes of the World War.

There is one ray of hope. I believe that today the responsible leaders of the nations do, in the main, honestly desire to abolish war. The resistance to this absolutely necessary step arises from those unfortunate national traditions which are handed on like a hereditary disease from generation to generation through the workings of the educational system. But the principal vehicle of this tradition is military training and its glorification, and, equally, that portion of the Press which is controlled by heavy industry and the military. Without disarmament there can be no lasting peace. Conversely, the continuation of the armament race on the present scale will inevitably lead to new catastrophes.

That is why the Disarmament Conference of 1932 will decide the fate of this generation and the next. When one thinks how pitiable, on the whole, have been the results of former conferences, it becomes clear that it is the duty of all intelligent and responsible people to exert their full powers to remind public opinion again and again of the importance of the 1932 Conference. Only if the statesmen have behind them the will to peace of a decisive majority in their own countries can they attain their great end, and for the formation of this public opinion each one of us is responsible in every word and deed.

The doom of the Conference would be sealed if the delegates came to it with ready-made instructions for a policy: to impose it on the Conference would at once become a matter of prestige. This seems to be generally realized. For meetings between the statesmen of two nations at a time, which have become very frequent of late, have been used to prepare the ground for the Conference by conversations about the disarmament problem. This seems to me a very happy device, for two men or groups of men can usually discuss things together most reasonably, honestly, and dispassionately when there is no third person present in front of whom they think they must be careful what they say. Only if exhaustive preparations of this kind are made for the Conference, if surprises are thereby ruled out, and if an atmosphere of confidence is created by genuine good will, can we hope for a happy issue.

In these great matters success is not a matter of cleverness, still less of cunning, but of honesty and confidence. The moral element cannot be displaced by reason, thank heaven, I am inclined to say.

The individual must not merely wait and criticize. He must serve the cause as best he can. The fate of the world will be such as the world deserves.

AMERICA AND THE DISARMAMENT
CONFERENCE OF 1932

Mein Weltbild, *Amsterdam: Querido Verlag, 1934.*

The Americans of today are filled with the cares arising out of the economic conditions in their own country. The efforts of their responsible leaders are directed primarily to remedying the serious unemployment at home. The sense of being involved in the destiny of the rest of the world, and in particular of the mother country of Europe, is even less strong than in normal times.

But the free play of economic forces will not by itself automatically overcome these difficulties. Regulative measures by the community are needed to bring about a sound distribution of labor and consumers' goods among mankind; without this even the people of the richest country suffocate. The fact is that since the amount of work needed to supply everybody's needs has been reduced through the improvement of technical methods, the free play of economic forces no longer produces a state of affairs in which all the available labor can find employment. Deliberate regulation and organization are becoming necessary to make the results of technical progress beneficial to all.

If the economic situation cannot be cleared up without systematic regulation, how much more necessary is such regulation for dealing with the international problems of politics! Few of us still cling to the notion that acts of violence in the shape of wars are either advantageous or worthy of humanity as a method of solving international problems. But we are not consistent enough to make vigorous efforts on behalf of the measures which might prevent war, that savage and unworthy relic of the age of barbarism. It requires some power of reflection to see the issue clearly and a certain courage to serve this great cause resolutely and effectively.

Anybody who really wants to abolish war must resolutely declare himself in favor of his own country's resigning a portion of its sovereignty in favor of international institutions: he must be ready to make his own country amenable, in case of a dispute, to the award of an international court. He must, in the most uncompromising fashion, support disarmament all round, as is actually envisaged in the unfortunate Treaty of Versailles; unless military and aggressively patriotic education is abolished, we can hope for no progress.

No event of the last few years reflects such disgrace on the leading civilized countries of the world as the failure of all disarmament conferences so far; for this failure is due not only to the intrigues of ambitious and unscrupulous politicians but also to the indifference and slackness of the public in all countries. Unless this is changed we shall destroy all the really valuable achievements of our predecessors.

I believe that the American people are only imperfectly aware of the responsibility which rests with them in this matter.

They no doubt think "Let Europe go to the dogs, if she is destroyed by the quarrelsomeness and wickedness of her inhabitants. The good seed of our Wilson has produced a mighty poor crop in the stony group of Europe. We are strong and safe and in no hurry to mix ourselves up in other people's affairs."

Such an attitude is neither noble nor far-sighted. America is partly to blame for the difficulties of Europe. By ruthlessly pressing her claims she is hastening the economic and therewith the moral decline of Europe; she has helped to Balkanize Europe and therefore shares the responsibility for the breakdown of political morality and the growth of that spirit of revenge which feeds on despair. This spirit will not stop short of the gates of America—I had almost said, has not stopped short. Look around, and beware!

The truth can be briefly stated:—The Disarmament Conference comes as a final chance, to you no less than to us, of preserving the best that civilized humanity has produced. And it is on you, as the strongest and comparatively soundest among us, that the eyes and hopes of all are focused.

THE QUESTION OF DISARMAMENT

Mein Weltbild, *Amsterdam: Querido Verlag, 1934.*

The greatest obstacle to the success of the disarmament plan was the fact that people in general left out of account the chief difficulties of the problem. Most objects are gained by gradual steps: for example, the supersession of absolute monarchy by democracy. Here, however, we are concerned with an objective which cannot be reached step by step.

As long as the possibility of war remains, nations will insist on being as perfectly prepared in a military sense as they can, in order to emerge triumphant from the next war. It will also be impossible to avoid educating the youth in warlike traditions and cultivating narrow national vanity joined to the glorification of the warlike spirit, as long as people have to be prepared for occasions when such a spirit will be needed for the purpose of war. To arm is to give one's voice and make one's preparations, not for peace but for war. Therefore people will not disarm step by step; they will disarm at one blow or not at all.

The accomplishment of such a far-reaching change in the life of nations presupposes a mighty moral effort, a deliberate departure from deeply ingrained tradition. Anyone who is not prepared to make the fate of his country in case of a dispute depend entirely on the decisions of an international court of arbitration, and to enter into a treaty to this effect without reserve, is not really resolved to avoid war. It is a case of all or nothing.

It is undeniable that previous attempts to ensure peace have failed through aiming at inadequate compromises.

Disarmament and security are only to be had in combination. The one guarantee of security is an undertaking by all nations to give effect to the decisions of the international authority.

We stand, therefore, at the parting of the ways. Whether we find the way of peace or continue along the old road of brute force, so unworthy of our civilization, depends on ourselves. On the one side the freedom of the individual and the security of society beckon to us; on the other, slavery for the individual and the annihilation of our civilization threaten us. Our fate will be according to our deserts.

ARBITRATION

Mein Weltbild, *Amsterdam: Querido Verlag, 1934.*

Systematic disarmament within a short period is only possible in combination with a guarantee of all nations for the security of each individual one, based on a permanent court of arbitration independent of governments.

Unconditional obligation of all countries not merely to accept the decisions of the court of arbitration but also to carry them out.

Separate courts of arbitration for Europe with Africa, America, and Asia (Australia to be apportioned to one of these). A joint court of arbitration for questions involving issues that cannot be settled within the limits of any one of these three regions.

TO SIGMUND FREUD

A private letter written around 1931 or the beginning of 1932. Published in Mein Weltbild, *Amsterdam: Querido Verlag, 1934.*

DEAR PROFESSOR FREUD:

It is admirable how the yearning to perceive the truth has overcome every other yearning in you. You have shown with impelling lucidity how inseparably the combative and destructive instincts are bound up in the human psyche with those of love and life. But at the same time there shines through the cogent logic of your arguments a deep longing for the great goal of internal and external liberation of mankind from war. This great aim has been professed by all those who have been venerated as moral and spiritual leaders beyond the limits of their own time and country without exception, from Jesus Christ to Goethe and Kant. Is it not significant that such men have been universally accepted as leaders, even though their efforts to mold the course of human affairs were attended with but small success?

I am convinced that the great men, those whose achievements in howsoever restricted a sphere set them above their fellows, share to an overwhelming extent the same ideal. But they have little influence on the course of political events. It almost looks as if this domain on which the fate of nations depends has inescapably to be given over to the violence and irresponsibility of political rulers.

Political leaders or governments owe their position partly to force and partly to popular election. They cannot be regarded as representative of the best elements, morally or intellectually, in their respective nations. The intellectual élite have no direct influence on the history of nations in these days; their lack of cohesion prevents them from taking a direct part in the solution of contemporary problems. Don't you think that a change might be brought about in this respect by a free association of people whose previous achievements and actions constitute a guarantee of their ability and purity of aim? This association of an international nature, whose members would need to keep in touch with each other by a constant interchange of opinions, might, by defining its attitude in the Press—responsibility always resting with the signatories on any given occasion—acquire a considerable and salutary moral influence over the settlement of political questions. Such an association would, of course, be a prey to all the ills which so often lead to degeneration in learned societies, dangers which are inseparably bound up with the imperfections of human nature. But should not an effort in this direction be risked in spite of this? I look upon such an attempt as nothing less than an imperative duty.

If an intellectual association of standing, such as I have described, could be formed, it would also have to make a consistent effort to mobilize the religious organizations for the fight against war. It would give countenance to many whose good intentions are paralyzed today by a melancholy resignation. Finally, I believe that an association formed of persons such as I have described, each highly esteemed in his own line, would be well suited to give valuable moral support to those elements in the League of Nations which are really working toward the great objective for which that institution exists.

I had rather put these proposals to you than to anyone else in the world, because you, least of all men, are the dupe of your desires and because your critical judgment is supported by a most grave sense of responsibility.

PEACE

*Since the time this article was written, it has been generally
recognized that the view expressed here, which prevailed
in the 1930's, is too narrow an interpretation of causes.
Nevertheless the conclusion still holds true. Published in*
Mein Weltbild, *Amsterdam: Querido Verlag, 1934.*

The importance of securing international peace was recognized by the really great men of former generations. But the
technical advances of our times have turned this ethical postulate into a matter of life and death for civilized mankind today,
and made it a moral duty to take an active part in the solution
of the problem of peace, a duty which no conscientious man
can shirk.

One has to realize that the powerful industrial groups concerned in the manufacture of arms are doing their best in all
countries to prevent the peaceful settlement of international
disputes, and that rulers can only achieve this great end if they
are sure of the vigorous support of the majority of their people.
In these days of democratic government the fate of nations hangs
on the people themselves; each individual must always bear
that in mind.

THE PACIFIST PROBLEM

Mein Weltbild, *Amsterdam: Querido Verlag, 1934.*

LADIES AND GENTLEMEN:

I am very glad of this opportunity of saying a few words to
you about the problem of pacifism. The course of events in the
last few years has once more shown us how little we are justified
in leaving the struggle against armaments and against the war
spirit to the governments. On the other hand, the formation
of large organizations with a large membership can in itself
bring us very little nearer to our goal. In my opinion, the best
method in this case is the violent one—conscientious objection,

which must be aided by organizations that give moral and material support to the courageous conscientious objectors in each country. In this way we may succeed in making the problem of pacifism an acute one, a real struggle to which forceful spirits will be attracted. It is an illegal struggle, but a struggle for the true rights of the people against their governments as far as they demand criminal acts of their citizens.

Many who think themselves good pacifists will jibe at this out and out pacifism, on patriotic grounds. Such people are not to be relied on in the hour of crisis, as the World War amply proved.

I am most grateful to you for according me an opportunity to give you my views in person.

COMPULSORY SERVICE

Mein Weltbild, *Amsterdam: Querido Verlag, 1934.*

Instead of permission being given to Germany to introduce compulsory service, it ought to be taken away from all other powers: to begin with none but mercenary armies should be permitted, the size and equipment of which should be discussed at Geneva. This would also be better for France than to be forced to permit compulsory service in Germany. The fatal psychological effect of the military education of the people and the violation of the individual's rights which it involves would thus be avoided.

Moreover, it would be much easier for two countries agreeing to compulsory arbitration for the settlement of all disputes concerning their mutual relations to combine such mercenary forces into a single organization with mixed units. This would mean financial relief and increased security for both of them. Such a process of amalgamation might extend to larger and larger combinations, and finally lead to an international police, which would have to decay gradually with the increase of international security.

Will you discuss this proposal with our friends by way of

setting the ball rolling? Of course I do not in the least insist on this particular proposal. But I do think it essential that we should came forward with a positive program; a merely negative policy is unlikely to produce any practical results.

WOMEN AND WAR

Retort to American women. The "defenseless civilian" is Albert Einstein. Published in Mein Weltbild, *Amsterdam: Querido Verlag, 1934.*

In my opinion, the patriotic women ought to be sent to the front in the next war instead of the men. It would at least be a novelty in this dreary sphere of infinite confusion. And besides, why should not such heroic feelings on the part of the fair sex find a more picturesque outlet than in attacks on a defenseless civilian?

THREE LETTERS TO FRIENDS OF PEACE

Mein Weltbild, *Amsterdam: Querido Verlag, 1934.*

I.

It has come to my knowledge that out of the greatness of your soul you are quietly accomplishing a splendid work, impelled by solicitude for humanity and its fate. Small is the number of them that see with their own eyes and feel with their own hearts. But it is their strength that will decide whether the human race must relapse into that state of stupor which a deluded multitude appears today to regard as the ideal.

O that the nations might see, before it is too late, how much of their self-determination they have got to sacrifice in order to avoid the struggle of all against all! The power of conscience and of the international spirit has proved itself inadequate. At present it is being so weak as to tolerate parleying with the worst enemies of civilization. There is a kind of compliance which is a crime against humanity, though it passes for political wisdom.

We cannot despair of humanity, since we are ourselves human beings. And it is a comfort that there still exist individuals like yourself, whom one knows to be alive and undismayed.

II.

To be quite frank, a declaration like the one before me in a country which submits to conscription in peace-times seems to me worthless. What you must fight for is liberation from universal military service. Verily, the French nation has had to pay heavily for the victory of 1918; for that victory has been largely responsible for holding it down in the most degrading of all forms of slavery.

Let your efforts in this struggle be unceasing. You have a mighty ally in the German reactionaries and militarists. If France clings to universal military service, it will be impossible in the long run to prevent its introduction into Germany. For the demand of the Germans for equal rights will succeed in the end; and then there will be two German military slaves to every French one, which would certainly not be in the interests of France.

Only if we succeed in abolishing compulsory service altogether will it be possible to educate the youth in the spirit of reconciliation, joy in life, and love toward all living creatures.

I believe that a refusal on conscientious grounds to serve in the army when called up, if carried out by 50,000 men at the same moment, would be irresistible. The individual can accomplish little here, nor can one wish to see the best among us devoted to destruction at the hands of the machinery behind which stand three great powers: stupidity, fear, and greed.

III.

The point with which you deal in your letter is one of prime importance. The armament industry is indeed one of the

greatest dangers that beset mankind. It is the hidden evil power
behind the nationalism which is rampant everywhere. . . .

Possibly something might be gained by nationalization. But
it is extremely hard to determine exactly what industries should
be included. Should the aircraft industry? And how much of
the metal industry and the chemical industry?

As regards the munitions industry and the export of war
material, the League of Nations has busied itself for years with
efforts to get this loathsome traffic controlled—with what little
success, we all know. Last year I asked a well-known American
diplomat why Japan was not forced by a commercial boycott to
desist from her policy of force. "Our commercial interests are
too strong" was the answer. How can one help people who
rest satisfied with a statement like that?

You believe that a word from me would suffice to get something done in this sphere? What an illusion! People flatter
me as long as I do not get in their way. But if I direct my
efforts toward objects which do not suit them, they immediately
turn to abuse and calumny in defense of their interests. And
the onlookers mostly keep out of the limelight, the cowards!
Have you ever tested the civil courage of your countrymen?
The silently accepted motto is "Leave it alone and say nothing
about it." You may be sure that I shall do everything in my
power along the lines you indicate, but nothing can be achieved
as directly as you think.

ACTIVE PACIFISM

Mein Weltbild, *Amsterdam: Querido Verlag, 1934.*

I consider myself lucky to have witnessed the great peace
demonstration which the Flemish people has undertaken. To
all concerned in it I feel impelled to call out, in the name of all
men of good will who care for the future: "In this hour of
reflection and awakening of the conscience we feel deeply united
with you."

We must not conceal from ourselves that no improvement in the present depressing situation is possible without a severe struggle; for the handful of those who are really determined to do something is minute in comparison with the mass of the lukewarm and the misguided. And those who have an interest in keeping the machinery of war going are a very powerful body; they will stop at nothing to make public opinion subservient to their murderous ends.

It looks as if the ruling statesmen of today were really trying to secure permanent peace. But the ceaseless piling-up of armaments shows only too clearly that they are unequal to coping with the hostile forces which are preparing for war. In my opinion, deliverance can only come from the peoples themselves. If they wish to avoid the degrading slavery of war-service, they must declare with no uncertain voice for complete disarmament. As long as armies exist, any serious conflict will lead to war. A pacifism which does not actively fight against the armament of nations is and must remain impotent.

May the conscience and the common sense of the peoples be awakened, so that we may reach a new stage in the life of nations, where people will look back on war as an incomprehensible aberration of their forefathers!

OBSERVATIONS ON THE PRESENT SITUATION IN EUROPE

Mein Weltbild, *Amsterdam: Querido Verlag, 1934.*

The distinguishing feature of the present political situation of the world, and in particular of Europe, seems to me to be this, that political development has failed, both materially and intellectually, to keep pace with economic necessity, which has changed its character in a comparatively short time. The interests of each country must be subordinated to the interests of the wider community. The struggle for this new orientation of political thought and feeling is a severe one, because it has the tradition of centuries against it. But the survival of Europe depends on its successful issue. It is my firm conviction that once the psychological impediments are overcome, the solution of the real problems will not be such a terribly difficult matter. In order to create the right atmosphere, the most essential thing is personal cooperation between men of like mind. May our united efforts succeed in building a bridge of mutual trust between the nations!

GERMANY AND FRANCE

Mein Weltbild, *Amsterdam: Querido Verlag, 1934.*

Mutual trust and cooperation between France and Germany can only come about if the French demand for security against military attack is satisfied. But should France frame demands in accordance with this, such a step would certainly be taken very ill in Germany.

A procedure like the following seems, however, to be possible. Let the German government of its own free will propose to the French that they should jointly make representations to the League of Nations that it should suggest to all member states to bind themselves to the following:—

(1) To submit to every decision of the international court of arbitration.

(2) To proceed with all its economic and military force, in concert with the other members of the League, against any state which breaks the peace or resists an international decision made in the interests of world peace.

CULTURE AND PROSPERITY

Mein Weltbild, *Amsterdam: Querido Verlag, 1934.*

If one would estimate the damage done by the great political catastrophe to the development of human civilization, one must remember that culture in its higher forms is a delicate plant which depends on a complicated set of conditions and is wont to flourish only in a few places at any given time. For it to blossom there is needed, first of all, a certain degree of prosperity which enables a fraction of the population to work at things not directly necessary to the maintenance of life; second, a moral tradition of respect for cultural values and achievements, in virtue of which this class is provided with the means of living by the other classes, those who provide the immediate necessities of life.

During the past century Germany has been one of the countries in which both conditions were fulfilled. The prosperity was, taken as a whole, modest but sufficient; the tradition of respect for culture, vigorous. On this basis the German nation has brought forth fruits of culture which form an integral part of the development of the modern world. The tradition, in the main, still stands, though the prosperity is gone. The industries of the country have been cut off almost completely from the sources of raw materials on which the existence of the industrial part of the population was based. The surplus necessary to support the intellectual worker has suddenly ceased to exist. With it the tradition which depends on it will inevitably collapse also, and a fruitful nursery of culture turn to wilderness.

The human race, in so far as it sets a value on culture, has an interest in preventing such impoverishment. It will give what help it can in the immediate crisis and reawaken that higher community of feeling, now thrust into the background by national egotism, for which human values have a validity independent of politics and frontiers. It will then procure for every nation conditions of work under which it can exist and under which it can bring forth fruits of culture.

MINORITIES

Mein Weltbild, *Amsterdam: Querido Verlag, 1934.*

It seems to be a universal fact that minorities—especially when the individuals composing them can be recognized by physical characteristics—are treated by the majorities among whom they live as an inferior order of beings. The tragedy of such a fate lies not merely in the unfair treatment to which these minorities are automatically subjected in social and economic matters, but also in the fact that under the suggestive influence of the majority most of the victims themselves succumb to the same prejudice and regard their kind as inferior beings. This second and greater part of the evil can be overcome by closer association and by deliberate education of the minority, whose spiritual liberation can thus be accomplished.

The resolute efforts of the American Negroes in this direction deserve approval and assistance.

THE HEIRS OF THE AGES

Mein Weltbild, *Amsterdam: Querido Verlag, 1934.*

Previous generations were able to look upon intellectual and cultural progress as simply the inherited fruits of their forebears' labors, which made life easier and more beautiful for them. But the calamities of our times show us that this was a fatal illusion.

We see now that the greatest efforts are needed if this legacy of humanity's is to prove a blessing and not a curse. For whereas formerly it was enough for a man to have freed himself to some extent from personal egotism to make him a valuable member of society, today he must also be required to overcome national and class egotism. Only if he reaches those heights can he contribute toward improving the lot of humanity.

As regards this most important need of the age, the inhabitants of a small state are better placed than those of a great power, since the latter are exposed, both in politics and economics, to the temptation to gain their ends by brute force. The agreement between Holland and Belgium, which is the only bright spot in European events during the last few years, encourages one to hope that the small nations will play a leading part in the attempt to liberate the world from the degrading yoke of militarism through the renunciation of the individual country's unlimited right of self-determination.

THE WAR IS WON, BUT THE PEACE IS NOT

From an address on the occasion of the Fifth Nobel Anniversary Dinner at the Hotel Astor in New York, December 10, 1945. Published in Out of My Later Years, *New York: Philosophical Library, 1950.*

Physicists find themselves in a position not unlike that of Alfred Nobel. Alfred Nobel invented the most powerful explosive ever known up to his time, a means of destruction par excellence. In order to atone for this, in order to relieve his human conscience, he instituted his awards for the promotion of peace and for achievements of peace. Today, the physicists who participated in forging the most formidable and dangerous weapon of all times are harassed by an equal feeling of responsibility, not to say guilt. And we cannot desist from warning, and warning again, we cannot and should not slacken in our efforts to make the nations of the world, and especially their governments, aware of the unspeakable disaster they are certain to provoke unless they change their attitude toward each other and toward the task of shaping the future. We helped in creating this new weapon in order to prevent the enemies of mankind from achieving it ahead of us, which, given the mentality of the Nazis, would have meant inconceivable destruction and the enslavement of the rest of the world. We delivered this weapon into the hands of the American and the British people as trustees of the whole of mankind, as fighters for peace and liberty. But so far we fail to see any guarantee of peace, we do not see any guarantee of the freedoms that were promised to the nations in the Atlantic Charter. The war is won, but the peace is not. The great powers, united in fighting, are now divided over the peace settlements. The world was promised freedom from fear, but in fact fear has increased tremendously since the termination of the war. The world was promised freedom from want, but large parts of the world are faced with starvation while others are living in abundance. The nations were promised liberation and justice. But we have witnessed, and are witnessing even now, the sad spectacle of "liberating" armies firing into populations who want their independence and social equality, and supporting in those countries, by force of arms, such parties and personalities as appear to be most suited to serve vested interests. Territorial questions and arguments of power, obsolete though they are, still prevail over the essential demands of common welfare and justice. Allow me to be more specific about just one case, which is but a symptom of the general situation: the case of my own people, the Jewish people.

So long as Nazi violence was unleashed only, or mainly, against the Jews, the rest of the world looked on passively, and even treaties and agreements were made with the patently criminal government of the Third Reich. Later, when Hitler was on the point of taking over Rumania and Hungary, at the time when Maidanek and Oswiecim were in Allied hands, and the methods of the gas chambers were well known all over the world, all attempts to rescue the Rumanian and Hungarian Jews came to naught because the doors of Palestine were closed to Jewish immigrants by the British government, and no country could be found that would admit those forsaken people. They were left to perish like their brothers and sisters in the occupied countries.

We shall never forget the heroic efforts of the small countries, of the Scandinavian, the Dutch, the Swiss nations, and of individuals in the occupied parts of Europe who did all in their power to protect Jewish lives. We do not forget the humane attitude of the Soviet Union who was the only one among the big powers to open her doors to hundreds of thousands of Jews when the Nazi armies were advancing in Poland. But after all that has happened, and was not prevented from happening, how is it today? While in Europe territories are being distributed without any qualms about the wishes of the people concerned, the remainders of European Jewry, one-fifth of its prewar population, are again denied access to their haven in Palestine and left to hunger and cold and persisting hostility. There is no country, even today, that would be willing or able to offer them a place where they could live in peace and security. And the fact that many of them are still kept in the degrading conditions of concentration camps by the Allies gives sufficient evidence of the shamefulness and hopelessness of the situation. These people are forbidden to enter Palestine with reference to the principle of democracy, but actually the Western powers, in upholding the ban of the White Paper, are yielding to the threats and the external pressure of five vast and underpopulated Arab States. It is sheer irony when the British Foreign Minister tells the poor lot of European Jews they should remain in Europe because their genius is needed there, and, on the other hand, advises them not to try to get at the head of the queue lest they might incur new hatred and persecution. Well, I am afraid they cannot help it; with their six million dead they have been pushed at the head of the queue, of the queue of Nazi victims, much against their will.

The picture of our postwar world is not bright. So far as we, the physicists, are concerned, we are no politicians and it has never been our wish to meddle in politics. But we know a few things that the politicians do not know. And we feel the duty to speak up and to remind those responsible that there is no escape into easy comforts, there is no distance ahead for proceeding little by little and delaying the necessary changes into an indefinite future, there is no time left for petty bargaining. The situation calls for a courageous effort, for a radical change in our whole attitude, in the entire political concept. May the spirit that prompted Alfred Nobel to create his great institution, the spirit of trust and confidence, of generosity and brotherhood among men, prevail in the minds of those upon whose decisions our destiny rests. Otherwise, human civilization will be doomed.

ATOMIC WAR OR PEACE

From Atlantic Monthly, *Boston, November, 1945, and November, 1947. As told to Raymond Swing.*

I.

The release of atomic energy has not created a new problem. It has merely made more urgent the necessity of solving an existing one. One could say that it has affected us quantitatively, not qualitatively. So long as there are sovereign nations possessing great power, war is inevitable. That is not an attempt to say when it will come, but only that it is sure to come. That was true before the atomic bomb was made. What has been changed is the destructiveness of war.

I do not believe that civilization will be wiped out in a war fought with the atomic bomb. Perhaps two-thirds of the people of the earth might be killed. But enough men capable of thinking, and enough books, would be left to start again, and civilization could be restored.

I do not believe that the secret of the bomb should be given to the United Nations Organization. I do not believe it should be given to the Soviet Union. Either course would be like a man with capital, and wishing another man to work with him on some enterprise, starting out by simply giving that man half of his money. The other man might choose to start a rival enterprise, when what is wanted is his cooperation. The secret of the bomb should be committed to a world government, and the United States should immediately announce its readiness to give it to a world government. This government should be founded by the United States, the Soviet Union, and Great Britain, the only three powers with great military strength. All three of them should commit to this world government all of their military strength. The fact that there are only three nations with great military power should make it easier, rather than harder, to establish such a government.

Since the United States and Great Britain have the secret of the atomic bomb and the Soviet Union does not, they should invite the Soviet Union to prepare and present the first draft of a constitution of the proposed world government. That will help dispel the distrust of the Russians, which they already feel because the bomb is being kept a secret chiefly to prevent their having it. Obviously the first draft would not be the final one, but the Russians should be made to feel that the world government will assure them their security.

It would be wise if this constitution were to be negotiated by a single American, a single Briton, and a single Russian. They would have to have advisers, but these advisers should only advise when asked. I believe three men can succeed in writing a workable constitution acceptable to them all. Six or seven men, or more, probably would fail. After the three great powers have drafted a constitution and adopted it, the smaller nations should be invited to join the world government. They should be free to stay out, and though they should feel perfectly secure in staying out, I am sure they would wish to join. Naturally they should be entitled to propose changes in the constitution as drafted by the Big Three. But the Big Three should go ahead and organize the world government, whether the smaller nations join or not.

The power of this world government would be over all military matters, and there need be only one further power. That is to interfere in countries where a minority is oppressing a majority, and so is creating the kind of instability that leads to war. Conditions such as exist in Argentina and Spain should be dealt with. There must be an end to the concept of non-intervention, for to end it is part of keeping the peace.

The establishment of this world government must not have to wait until the same conditions of freedom are to be found in all three of the great powers. While it is true that in the Soviet Union the minority rules, I do not consider that internal conditions there are of themselves a threat to world peace. One must bear in mind that the people in Russia did not have a long political education, and changes to improve

Russian conditions had to be carried through by a minority for the reason that there was no majority capable of doing it. If I had been born a Russian, I believe I could have adjusted myself to this situation.

It should not be necessary, in establishing a world government with a monopoly of military authority, to change the structure of the three great powers. It would be for the three individuals who draft the constitution to devise ways for their different structures to be fitted together for collaboration.

Do I fear the tyranny of a world government? Of course I do. But I fear still more the coming of another war or wars. Any government is certain to be evil to some extent. But a world government is preferable to the far greater evil of wars, particularly with their intensified destructiveness. If such a world government is not established by a process of agreement, I believe it will come anyway, and in a much more dangerous form. For war or wars will end in one power being supreme and dominating the rest of the world by its overwhelming military strength.

Now we have the atomic secret, we must not lose it, and that is what we should risk doing, if we give it to the United Nations Organization or to the Soviet Union. But we must make it clear as quickly as possible that we are not keeping the bomb a secret for the sake of our power, but in the hope of establishing peace through a world government, and we will do our utmost to bring this world government into being.

I appreciate that there are persons who favor a gradual approach to world government, even though they approve of it as the ultimate objective. The trouble with taking little steps, one at a time, in the hope of reaching the ultimate goal, is that while they are being taken, we continue to keep the bomb without making our reason convincing to those who do not have it. That of itself creates fear and suspicion, with the consequence that the relations of rival sovereignties deteriorate dangerously. So while persons who take only a step at a time may think they are approaching world peace, they actually are contributing by their slow pace to the coming of war. We have no time to spend in this way. If war is to be averted, it must be done quickly.

We shall not have the secret very long. I know it is argued that no other country has money enough to spend on the development of the atomic bomb, which assures us the secret for a long time. It is a mistake often made in this country to measure things by the amount of money they cost. But other countries which have the materials and the men and care to apply them to the work of developing atomic power can do so, for men and materials and the decision to use them, and not money, are all that are needed.

I do not consider myself the father of the release of atomic energy. My part in it was quite indirect. I did not, in fact, foresee that it would be released in my time. I believed only that it was theoretically possible. It became practical through the accidental discovery of chain reaction, and this was not something I could have predicted. It was discovered by Hahn in Berlin, and he himself misinterpreted what he discovered. It was Lise Meitner who provided the correct interpretation, and escaped from Germany to place the information in the hands of Niels Bohr.

I do not believe that a great era of atomic science is to be assured by organizing science, in the way large corporations are organized. One can organize to apply a discovery already made, but not to make one. Only a free individual can make a discovery. There can be a kind of organizing by which scientists are assured their freedom and proper conditions of work. Professors of science in American universities, for instance, should be relieved of some of their teaching so as to have time for more research. Can you imagine an organization of scientists making the discoveries of Charles Darwin?

Nor do I believe that the vast private corporations of the United States are suitable to the needs of these times. If a visitor should come to this country from another planet, would he not find it strange that in this country so much power is permitted to private corporations without their having commensurate responsibility? I say this to stress that the American government must keep the control of atomic energy, not because socialism is necessarily desirable, but because atomic energy was developed by the government, and it would be unthinkable to turn over this property of the people to any individuals or groups of individuals. As to socialism, unless it is international to the extent of producing world government which controls all military power, it might more easily lead to wars than capitalism, because it represents a still greater concentration of power.

To give any estimate when atomic energy can be applied to constructive purposes is impossible. What now is known is only how to use a fairly large quantity of uranium. The use of small quantities, sufficient, say, to operate a car or an' airplane, so far is impossible, and one cannot predict when it will be achieved. No doubt, it will be achieved, but nobody can say when. Nor can one predict when materials more common than uranium can be used to supply atomic energy. Presumably all materials used for this purpose will be among the heavier elements of high atomic weight. Those elements are relatively scarce due to their lesser stability. Most of these materials may have already disappeared by radioactive disintegration. So though the release of atomic energy can be, and no doubt will be, a great boon to mankind, that may not be for some time.

I myself do not have the gift of explanation with which I am able to persuade large numbers of people of the urgency of the problems the human race now faces. Hence I should like to commend someone who has this gift of explanation, Emory Reves, whose book, *The Anatomy of the Peace,* is intelligent, clear, brief, and, if I may use the abused term, dynamic on the topic of war and need for world government.

Since I do not foresee that atomic energy is to be a great boon for a long time, I have to say that for the present it is a menace. Perhaps it is well that it should be. It may intimidate the human race to bring order into its international affairs,

which, without the pressure of fear, it undoubtedly would
not do.

II.

Since the completion of the first atomic bomb nothing has
been accomplished to make the world more safe from war,
while much has been done to increase the destructiveness of
war. I am not able to speak from any firsthand knowledge
about the development of the atomic bomb, since I do not
work in this field. But enough has been said by those who
do to indicate that the bomb has been made more effective.
Certainly the possibility can be envisaged of building a bomb
of far greater size, capable of producing destruction over a
larger area. It also is credible that an extensive use could be
made of radioactivated gases which would spread over a
wide region, causing heavy loss of life without damage to
buildings.

I do not believe it is necessary to go on beyond these possibilities to contemplate a vast extension of bacteriological
warfare. I am skeptical that this form presents dangers comparable with those of atomic warfare. Nor do I take into account a danger of starting a chain reaction of a scope great
enough to destroy part or all of this planet. I dismiss this on
the ground that if it could happen from a man-made atomic
explosion it would already have happened from the action
of the cosmic rays which are continually reaching the earth's
surface.

But it is not necessary to imagine the earth being destroyed like a nova by a stellar explosion to understand
vividly the growing scope of atomic war and to recognize that
unless another war is prevented it is likely to bring destruction on a scale never before held possible and even now
hardly conceived, and that little civilization would survive it.

In the first two years of the atomic era another phenomenon is to be noted. The public, having been warned of
the horrible nature of atomic warfare, has done nothing about

it, and to a large extent has dismissed the warning from its consciousness. A danger that cannot be averted had perhaps better be forgotten; or a danger against which every possible precaution has been taken also had probably better be forgotten. That is, if the United States had dispersed its industries and decentralized its cities, it might be reasonable for people to forget the peril they face.

I should say parenthetically that it is well that this country has not taken these precautions, for to have done so would make atomic war still more probable, since it would convince the rest of the world that we are resigned to it and are preparing for it. But nothing has been done to avert war, while much has been done to make atomic war more horrible; so there is no excuse for ignoring the danger.

I say that nothing has been done to avert war since the completion of the atomic bomb, despite the proposal for supranational control of atomic energy put forward by the United States in the United Nations. This country has made only a conditional proposal, and on conditions which the Soviet Union is now determined not to accept. This makes it possible to blame the failure on the Russians.

But in blaming the Russians the Americans should not ignore the fact that they themselves have not voluntarily renounced the use of the bomb as an ordinary weapon in the time before the achievement of supranational control, or if supranational control is not achieved. Thus they have fed the fear of other countries that they consider the bomb a legitimate part of their arsenal so long as other countries decline to accept their terms for supranational control.

Americans may be convinced of their determination not to launch an aggressive or preventive war. So they may believe it is superfluous to announce publicly that they will not a second time be the first to use the atomic bomb. But this country has been solemnly invited to renounce the use of the bomb—that is, to outlaw it—and has declined to do so unless its terms for supranational control are accepted.

I believe this policy is a mistake. I see a certain military gain from not renouncing the use of the bomb in that this may be deemed to restrain another country from starting a war in which the United States might use it. But what is gained in one way is lost in another. For an understanding over the supranational control of atomic energy has been made more remote. That may be no military drawback so long as the United States has the exclusive use of the bomb. But the moment another country is able to make it in substantial quantities, the United States loses greatly through the absence of an international agreement, because of the vulnerability of its concentrated industries and its highly developed urban life.

In refusing to outlaw the bomb while having the monopoly of it, this country suffers in another respect, in that it fails to return publicly to the ethical standards of warfare formally accepted previous to the last war. It should not be forgotten that the atomic bomb was made in this country as a preventive measure; it was to head off its use by the Germans, if they discovered it. The bombing of civilian centers was initiated by the Germans and adopted by the Japanese. To it the Allies responded in kind—as it turned out, with greater effectiveness—and they were morally justified in doing so. But now, without any provocation, and without the justification of reprisal or retaliation, a refusal to outlaw the use of the bomb save in reprisal is making a political purpose of its possession. This is hardly pardonable.

I am not saying that the United States should not manufacture and stockpile the bomb, for I believe that it must do so; it must be able to deter another nation from making an atomic attack when it also has the bomb. But deterrence should be the only purpose of the stockpile of bombs. In the same way I believe that the United Nations should have the atomic bomb when it is supplied with its own armed forces and weapons. But it, too, should have the bomb for the sole purpose of deterring an aggressor or rebellious nations from making an atomic attack. It should not use the atomic bomb on its own initiative any more than the United States or any other power should do so. To keep a stockpile of atomic bombs without promising not to initiate its use is exploiting the possession of the bombs for political ends. It may be that the United States hopes in this way to frighten the Soviet Union into accepting supranational control of atomic energy. But the creation of fear only heightens antagonism and increases the danger of war. I am of the opinion that this policy has detracted from the very real virtue in the offer of supranational control of atomic energy.

We have emerged from a war in which we had to accept the degradingly low ethical standards of the enemy. But instead of feeling liberated from his standards, and set free to restore the sanctity of human life and the safety of noncombatants, we are in effect making the low standards of the enemy in the last war our own for the present. Thus we are starting toward another war degraded by our own choice.

It may be that the public is not fully aware that in another war atomic bombs will be available in large quantities. It may measure the dangers in the terms of the three bombs exploded before the end of the last war. The public also may not appreciate that, in relation to the damage inflicted, atomic bombs already have become the most economical form of destruction that can be used on the offensive. In another war the bombs will be plentiful and they will be comparatively cheap. Unless there is determination not to use them that is far stronger than can be noted today among American political and military leaders, and on the part of the public itself, atomic warfare will be hard to avoid. Unless Americans come to recognize that they are not stronger in the world because they have the bomb, but weaker because of their vulnerability to atomic attack, they are not likely to conduct their policy at Lake Success or in their relations with Russia in a spirit that furthers the arrival at an understanding.

But I do not suggest that the American failure to outlaw the use of the bomb except in retaliation is the only cause of the absence of an agreement with the Soviet Union over atomic control. The Russians have made it clear that they will do everything in their power to prevent a supranational regime from coming into existence. They not only reject it in the range of atomic energy: they reject it sharply on principle, and thus have spurned in advance any overture to join a limited world government.

Mr. Gromyko has rightly said that the essence of the American atomic proposal is that national sovereignty is not compatible with the atomic era. He declares that the Soviet Union cannot accept this thesis. The reasons he gives are obscure, for they quite obviously are pretexts. But what seems to be true is that the Soviet leaders believe they cannot preserve the social structure of the Soviet state in a supranational regime. The Soviet government is determined to maintain its present social structure, and the leaders of Russia, who hold their great power through the nature of that structure, will spare no effort to prevent a supranational regime from coming into existence, to control atomic energy or anything else.

The Russians may be partly right about the difficulty of retaining their present social structure in a supranational regime, though in time they may be brought to see that this is a far lesser loss than remaining isolated from a world of law. But at present they appear to be guided by their fears, and one must admit that the United States has made ample contributions to these fears, not only as to atomic energy but in many other respects. Indeed this country has conducted its Russian policy as though it were convinced that fear is the greatest of all diplomatic instruments.

That the Russians are striving to prevent the formation of a supranational security system is no reason why the rest of the world should not work to create one. It has been pointed out that the Russians have a way of resisting with all their arts what they do not wish to have happen; but once it happens, they can be flexible and accommodate themselves to it. So it would be well for the United States and other powers not to permit the Russians to veto an attempt to create supranational security. They can proceed with some hope that once the Russians see they cannot prevent such a regime they may join it.

So far the United States has shown no interest in preserving the security of the Soviet Union. It has been interested in its own security, which is characteristic of the competition which marks the conflict for power between sovereign states. But one cannot know in advance what would be the effect on Russian fears if the American people forced their leaders to pursue a policy of substituting law for the present anarchy of international relations. In a world of law, Russian security would be equal to our own, and for the American people to espouse this wholeheartedly, something that should be possible under the workings of democracy, might work a kind of miracle in Russian thinking.

At present the Russians have no evidence to convince them that the American people are not contentedly supporting a policy of military preparedness which they regard as a policy of deliberate intimidation. If they had evidences of a passionate desire by Americans to preserve peace in the one way it can be maintained, by a supranational regime of law, this would upset Russian calculations about the peril to Russian security in current trends of American thought. Not until a genuine, convincing offer is made to the Soviet Union, backed by an aroused American public, will one be entitled to say what the Russian response would be.

It may be that the first response would be to reject the world of law. But if from that moment it began to be clear to the Russians that such a world was coming into existence without them, and that their own security was being increased, their ideas necessarily would change.

I am in favor of inviting the Russians to join a world government authorized to provide security, and if they are unwilling to join, to proceed to establish supranational security without them. Let me admit quickly that I see great peril in such a course. If it is adopted it must be done in a way to make it utterly clear that the new regime is not a combination of power against Russia. It must be a combination that by its composite nature will greatly reduce the chances of war. It will be more diverse in its interests than any single state, thus less likely to resort to aggressive or preventive war. It will be larger, hence stronger than any single nation. It will be geographically much more extensive, and thus more difficult to defeat by military means. It will be dedicated to supranational security, and thus escape the emphasis on national supremacy which is so strong a factor in war.

If a supranational regime is set up without Russia, its service to peace will depend on the skill and sincerity with which it is done. Emphasis should always be apparent on the desire to have Russia take part. It must be clear to Russia, and no less so to the nations comprising the organization, that no penalty is incurred or implied because a nation declines to join. If the Russians do not join at the outset, they must be sure of a welcome when they do decide to join. Those who create the organization must understand that they are building with the final objective of obtaining Russian adherence.

These are abstractions, and it is not easy to outline the specific lines a partial world government must follow to induce the Russians to join. But two conditions are clear to me: the new organization must have no military secrets; and the Russians must be free to have observers at every session of the organization, where its new laws are drafted, discussed, and adopted, and where its policies are decided. That would destroy the great factory of secrecy where so many of the world's suspicions are manufactured.

It may affront the military-minded person to suggest a regime that does not maintain any military secrets. He has been taught to believe that secrets thus divulged would enable a war-minded nation to seek to conquer the earth. (As to the so-called secret of the atomic bomb, I am assuming the Russians will have this through their own efforts within a short time.) I grant there is a risk in not maintaining military secrets. If a sufficient number of nations have pooled their strength they can take this risk, for their security will be greatly increased. And it can be done with greater assurance because of the decrease of fear, suspicion, and distrust that will result. The tensions of the increasing likelihood of war in a world based on sovereignty would be replaced by the relaxation of the growing confidence in peace. In time this might so allure the Russian people that their leaders would mellow in their attitude toward the West.

Membership in a supranational security system should not, in my opinion, be based on any arbitrary democratic standards. The one requirement from all should be that the representatives to a supranational organization—assembly and council— must be elected by the people in each member country through a secret ballot. These representatives must represent the people rather than any government—which would enhance the pacific nature of the organization.

To require that other democratic criteria be met is, I believe, inadvisable. Democratic institutions and standards are the result of historic developments to an extent not always appreciated in the lands which enjoy them. Setting arbitrary standards sharpens the ideological differences between the Western and Soviet systems.

But it is not the ideological differences which now are pushing the world in the direction of war. Indeed, if all the Western nations were to adopt socialism, while maintaining their national sovereignty, it is quite likely that the conflict for power between East and West would continue. The passion expressed over the economic systems of the present seems to me quite irrational. Whether the economic life of America should be dominated by relatively few individuals, as it is, or these individuals should be controlled by the state, may be important, but it is not important enough to justify all the feelings that are stirred up over it.

I should wish to see all the nations forming the supranational state pool all their military forces, keeping for themselves only local police. Then I should like to see these forces commingled and distributed as were the regiments of the former Austro-Hungarian Empire. There it was appreciated that the men and officers of one region would serve the purposes of empire better by not being stationed exclusively in their own provinces, subject to local and racial pulls.

I should like to see the authority of the supranational regime restricted altogether to the field of security. Whether this would be possible I am not sure. Experience may point to the desirability of adding some authority over economic matters, since under modern conditions these are capable of causing national upsets that have in them the seeds of violent conflict. But I should prefer to see the function of the organization altogether limited to the tasks of security. I also should like to see this regime established through the strengthening of the United Nations, so as not to sacrifice continuity in the search for peace.

I do not hide from myself the great difficulties of establishing a world government, either a beginning without Russia or one with Russia. I am aware of the risks. Since I should not wish it to be permissible for any country that has joined the supranational organizat on to secede, one of these risks is a possible civil war. But I also believe that world government is certain to come in time, and that the question is how much it is to be permitted to cost. It will come, I believe, even if there is another world war, though after such a war, if it is won, it would be world government established by the victor, resting on the victor's military power, and thus to be maintained permanently only through the permanent militarization of the human race.

But I also believe it can come through agreement and through the force of persuasion alone, hence at low cost. But if it is to come in this way, it will not be enough to appeal to reason. One strength of the communist system of the East is that it has some of the character of a religion and inspires the emotions of a religion. Unless the cause of peace based on law gathers behind it the force and zeal of a religion, it hardly can hope to succeed. Those to whom the moral teaching of the human race is entrusted surely have a great duty and a great opportunity. The atomic scientists, I think, have become convinced that they cannot arouse the American people to the truths of the atomic era by logic alone. There must be added that deep power of emotion which is a basic ingredient of religion. It is to be hoped that not only the churches but the schools, the colleges, and the leading organs of opinion will acquit themselves well of their unique responsibility in this regard.

THE MILITARY MENTALITY

From The American Scholar, *New York, Summer, 1947.*

It seems to me that the decisive point in the situation lies in the fact that the problem before us cannot be viewed as an isolated one. First of all, one may pose the following question: from now on institutions for learning and research will more and more have to be supported by grants from the state, since, for various reasons, private sources will not suffice. Is it at all reasonable that the distribution of the funds raised for these purposes from the taxpayer should be entrusted to the military? To this question every prudent person will certainly answer: "No!" For it is evident that the difficult task of the most beneficent distribution should be placed in the hands of people whose training and life's work give proof that they know something about science and scholarship.

If reasonable people, nevertheless, favor military agencies for the distribution of a major part of the available funds, the reason for this lies in the fact that they subordinate cultural concerns to their general political outlook. We must then focus our attention on these practical political viewpoints, their origins and their implications. In doing so we shall soon recognize that the problem here under discussion is but one of many, and can only be fully estimated and properly adjudged when placed in a broader framework.

The tendencies we have mentioned are something new for America. They arose when, under the influence of the two World Wars and the consequent concentration of all forces on a military goal, a predominantly military mentality developed, which with the almost sudden victory became even more accentuated. The characteristic feature of this mentality is that people place the importance of what Bertrand Russell so tellingly terms "naked power" far above all other factors which affect the relations between peoples. The Germans, misled by Bismarck's successes in particular, underwent just such a transformation of their mentality—in consequence of which they were entirely ruined in less than a hundred years.

I must frankly confess that the foreign policy of the United States since the termination of hostilities has reminded me, sometimes irresistibly, of the attitude of Germany under Kaiser Wilhelm II, and I know that, independent of me, this analogy has most painfully occurred to others as well. It is characteristic of the military mentality that non-human factors (atom bombs, strategic bases, weapons of all sorts, the possession of raw materials, etc.) are held essential, while the human being, his desires and thoughts—in short, the psychological factors—are considered as unimportant and secondary. Herein lies a certain resemblance to Marxism, at least in so far as its theoretical side alone is kept in view. The individual is degraded to a mere instrument; he becomes "human matériel." The normal ends of human aspiration vanish with such a viewpoint. Instead, the military mentality raises "naked power" as a goal in itself— one of the strangest illusions to which men can succumb.

In our time the military mentality is still more dangerous than formerly because the offensive weapons have become much more powerful than the defensive ones. Therefore it leads, by necessity, to preventive war. The general insecurity that goes hand in hand with this results in the sacrifice of the citizen's civil rights to the supposed welfare of the state. Political witch-hunting, controls of all sorts (e.g., control of teaching and research, of the press, and so forth) appear inevitable, and for this reason do not encounter that popular resistance, which, were it not for the military mentality, would provide a protection. A reappraisal of all values gradually takes place in so far as everything that does not clearly serve the utopian ends is regarded and treated as inferior.

I see no other way out of prevailing conditions than a far-seeing, honest, and courageous policy with the aim of establishing security on supranational foundations. Let us hope that men will be found, sufficient in number and moral force, to guide the nation on this path so long as a leading rôle is imposed on her by external circumstances. Then problems such as have been discussed here will cease to exist.

EXCHANGE OF LETTERS WITH MEMBERS OF THE RUSSIAN ACADEMY

From Moscow New Times, *November 26, 1947, and* Bulletin of the Atomic Scientists, *Chicago, February, 1948.*

AN OPEN LETTER: DR. EINSTEIN'S MISTAKEN NOTIONS

The celebrated physicist, Albert Einstein, is famed not only for his scientific discoveries; of late years he has paid much attention to social and political problems. He speaks over the radio and writes in the press. He is associated with a number of public organizations. Time and again he raised his voice in protest against the Nazi barbarians. He is an advocate of enduring peace, and has spoken against the threat of a new war, and against the ambition of the militarists to bring American science completely under their control.

Soviet scientists, and the Soviet people in general, are appreciative of the humanitarian spirit which prompts these activities of the scientist, although his position has not always been as consistent and clear-cut as might be desired. However, in some of Einstein's more recent utterances there have been aspects which seem to us not only mistaken, but positively prejudicial to the cause of peace which Einstein so warmly espouses.

We feel it our duty to draw attention to this, in order to clarify so important a question as to how most effectively to work for peace. It is from this point of view that the idea of a "world government" which Dr. Einstein has of late been sponsoring must be considered.

In the motley company of proponents of this idea, besides out-and-out imperialists who are using it as a screen for unlimited expansion, there are quite a number of intellectuals in the capitalist countries who are captivated by the plausibility of the idea, and who do not realize its actual implications. These pacifist and liberal-minded individuals believe that a "world government" would be an effective panacea against the world's evils and a guardian of enduring peace.

The advocates of a "world government" make wide use of the seemingly radical argument that in this atomic age state sovereignty is a relic of the past, that it is, as Spaak, the Belgian delegate, said in the UN General Assembly, an "old-fashioned" and even "reactionary" idea. It would be hard to imagine an allegation that is further from the truth.

In the first place, the idea of a "world government" and "superstate" are by no means products of the atomic age. They are much older than that. They were mooted, for instance, at the time the League of Nations was formed.

Further, these ideas have never been progressive in these modern times. They are a reflection of the fact that the capitalist monopolies, which dominate the major industrial countries, find their own national boundaries too narrow. They need a world-wide market, world-wide sources of raw materials, and world-wide spheres of capital investment. Thanks to their domination in political and administrative affairs, the monopoly interests of the big powers are in a position to utilize the machinery of government, in their struggle for spheres of influence and their efforts economically and politically to subjugate other countries, to play the master in these countries as freely as in their own.

We know this very well from the past experience of our own country. Under tsarism, Russia, with her reactionary regime, which was servilely accommodating to the interests of capital, with her low-paid labor and vast natural resources, was an alluring morsel to foreign capitalists. French, British, Belgian, and German firms battened on our country like birds of prey, earning profits which would have been inconceivable in their own countries. They chained tsarist Russia to the capitalist West with extortionate loans. Supported by funds obtained from foreign banks, the tsarist government brutally repressed the revolutionary movement, retarded the development of Russian science and culture, and instigated Jewish pogroms.

The Great October Socialist Revolution smashed the chains of economic and political dependence that bound our country to the world capitalist monopolies. The Soviet Government made our country for the first time a really free and independent state, promoted the progress of our Socialist economy, technology, science, and culture at a speed hitherto unwitnessed in history, and turned our country into a reliable bulwark of international peace and security. Our people upheld their country's independence in the civil war, in the struggle against the intervention of a bloc of imperialist states, and in the great battles of the war against the Nazi invaders.

And now the proponents of a "world superstate" are asking us voluntarily to surrender this independence for the sake of a "world government," which is nothing but a flamboyant signboard for the world supremacy of the capitalist monopolies.

It is obviously preposterous to ask of us anything like that. And it is not only with regard to the Soviet Union that such a demand is absurd. After World War II, a number of countries succeeded in breaking away from the imperialist system of oppression and slavery. The peoples of these countries are working to consolidate thei? economic and political independence, debarring alien interference in their domestic affairs. Further, the rapid spread of the movement for national independence in the colonies and dependencies has awakened the national consciousness of hundreds of millions of people, who do not desire to remain in the status of slaves any longer.

The monopolies of the imperialist countries, having lost a number of profitable spheres of exploitation, and running the risk of losing more, are doing their utmost to deprive the nations that have escaped from their mastery of the state independence which they, the monopolies, find so irksome, and to prevent the genuine liberation of the colonies. With this purpose, the imperialists are resorting to the most diverse methods of military, political, economic, and ideological warfare.

It is in accordance with this social behest that the ideologians of imperialism are endeavoring to discredit the very idea of national sovereignty. One of the methods they resort to is the advocacy of pretentious plans for a "world state," which will allegedly do away with imperialism, wars, and national enmity, ensure the triumph of universal law, and so on.

The predatory appetites of the imperialist forces that are striving for world supremacy are thus disguised under the garb of a pseudo-progressive idea which appeals to certain intellectuals—scientists, writers, and others—in the capitalist countries.

In an open letter which he addressed last September to the United Nations delegations, Dr. Einstein suggested a new scheme for limiting national sovereignty. He recommends that the General Assembly be reconstructed and converted into a permanently functioning world parliament endowed with greater authority than the Security Council, which, Einstein declares (repeating what the henchmen of American diplomacy are asserting day in and day out), is paralyzed by the veto right. The General Assembly, reconstructed in accordance with Dr. Einstein's plan, is to have final powers of decision, and the principle of the unanimity of the Great Powers is to be abandoned.

Einstein suggests that the delegates to the United Nations should be chosen by popular election and not appointed by their governments, as at present. At a first glance, this proposal may seem progressive and even radical. Actually, it will in no way improve the existing situation.

Let us picture to ourselves what elections to such a "world parliament" would mean in practice.

A large part of humanity still lives in colonial and dependent countries dominated by the governors, the troops, and the financial and industrial monopolies of a few imperialist powers. "Popular election" in such countries would in practice mean the appointment of delegates by the colonial administration or the military authorities. One does not have to go far for examples; one need only recall the parody of a referendum in Greece, which was carried out by her royalist-fascist rulers under the protection of British bayonets.

But things would be not much better in the countries where universal suffrage formally exists. In the bourgeois-democratic countries, where capital dominates. the latter resorts to thousands of tricks and devices to turn universal suffrage and freedom of ballot into a farce. Einstein surely knows that in the last Congressional elections in the United States only 39 per cent of the electorate went to the polls; he surely knows that millions of Negroes in the Southern states are virtually deprived of the franchise, or are forced, not infrequently under threat of lynching, to vote for their bitterest enemies, such as the late archreactionary and Negrophobe, Senator Bilbo.

Poll taxes, special tests, and other devices are employed to rob millions of immigrants, migrant workers, and poor farmers of the vote. We will not mention the widespread practice of purchasing votes, the rôle of the reactionary press, that powerful instrument for influencing the masses wielded by millionaire newspaper proprietors, and so forth.

All this shows what popular elections to a world parliament, as suggested by Einstein, would amount to under existing conditions in the capitalist world. Its composition would be no better than the present composition of the General Assembly. It would be a distorted reflection of the real sentiments of the masses, of their desire and hope for lasting peace.

As we know, in the General Assembly and the UN committees, the American delegation has a regular voting machine at its disposal, thanks to the fact that the overwhelming majority of the members of the UN are dependent on the United States and are compelled to adapt their foreign policy to the requirements of Washington. A number of Latin-American countries, for instance, countries with single-crop agricultural systems, are bound hand and foot to the American monopolies, which determine the prices of their produce. Such being the case, it is not surprising that, under pressure of the American delegation, a mechanical majority has arisen in the General Assembly which votes in obedience to the orders of its virtual masters.

There are cases when American diplomacy finds it preferable to realize certain measures, not through the State Department, but under the flag of the United Nations. Witness the notorious Balkan committee or the commission appointed to observe the elections in Korea. It is with the object of converting the UN into a branch of the State Department that the American delegation is forcing through the project for a "Little Assembly," which would in practice replace the Security Council, with its principle of unanimity of the Great Powers that is proving such an obstacle to the realization of imperialist schemes.

Einstein's suggestion would lead to the same result, and thus, far from promoting lasting peace and international cooperation, would only serve as a screen for an offensive against nations which have established regimes that prevent foreign capital from extorting its customary profits. It would further the unbridled expansion of American imperialism, and ideologically disarm the nations which insist upon maintaining their independence.

By the irony of fate, Einstein has virtually become a supporter of the schemes and ambitions of the bitterest foes of peace and international cooperation. He has gone so far in this direction as to declare in advance in his open letter that if the Soviet Union refuses to join his newfangled organization, other countries would have every right to go ahead without it, while leaving the door open for eventual Soviet participation in the organization as a member or as an "observer."

Essentially this proposal differs very little from the suggestions of frank advocates of American imperialism, however remote Dr. Einstein may be from them in reality. The sum and substance of these suggestions is that if UN cannot be converted into a weapon of United States policy, into a screen for imperialist schemes and designs, that organization should be wrecked and a new "international" organization formed in its place, without the Soviet Union and the new democracies.

Does Einstein not realize how fatal such plans would be to international security and international cooperation?

We believe that Dr. Einstein has entered a false and dangerous path; he is chasing the mirage of a "world state" in a world where different social, political, and economic systems exist. Of course there is no reason why states with different social and economic structures should not cooperate economically and politically, provided that these differences are soberly faced. But Einstein is sponsoring a political fad which plays into the hands of the sworn enemies of sincere international cooperation and enduring peace. The course he is inviting the member states of the United Nations to adopt would lead not to greater international security, but to new international complications. It would benefit only the capitalist monopolies, for whom new international complications hold out the promise of more war contracts and more profits.

It is because we so highly esteem Einstein as an eminent scientist and as a man of public spirit who is striving to the best of his ability to promote the cause of peace, that we consider it our duty to speak with utter frankness and without diplomatic adornment.

ALBERT EINSTEIN'S REPLY

Four of my Russian colleagues have published a benevolent attack upon me in an open letter carried by the *New Times*. I appreciate the effort they have made and I appreciate even more the fact that they have expressed their point of view so candidly and straightforwardly. To act intelligently in human affairs is only possible if an attempt is made to understand the thoughts, motives, and apprehensions of one's opponent so fully that one can see the world through his eyes. All well-meaning people should try to contribute as much as possible to improving such mutual understanding. It is in this spirit that I should like to ask my Russian colleagues and any other reader to accept the following answer to their letter. It is the reply of a man who anxiously tries to find a feasible solution without having the illusion that he himself knows "the truth" or "the right path" to follow. If in the following I shall express my views somewhat dogmatically, I do it only for the sake of clarity and simplicity.

Although your letter, in the main, is clothed in an attack

upon the non-socialistic foreign countries, particularly the United States, I believe that behind the aggressive front there lies a defensive mental attitude which is nothing else but the trend toward an almost unlimited isolationism. The escape into isolationism is not difficult to understand if one realizes what Russia has suffered at the hands of foreign countries during the last three decades—the German invasions with planned mass murder of the civilian population, foreign interventions during the civil war, the systematic campaign of calumnies in the western press, the support of Hitler as an alleged tool to fight Russia. However understandable this desire for isolation may be, it remains no less disastrous to Russia and to all other nations; I shall say more about it later on.

The chief object of your attack against me concerns my support of "world government." I should like to discuss this important problem only after having said a few words about the antagonism between socialism and capitalism; for your attitude on the significance of this antagonism seems to dominate completely your views on international problems. If the socio-economic problem is considered objectively, it appears as follows: technological development has led to increasing centralization of the economic mechanism. It is this development which is also responsible for the fact that economic power in all widely industrialized countries has become concentrated in the hands of relatively few. These people, in capitalist countries, do not need to account for their actions to the public as a whole; they must do so in socialist countries in which they are civil servants similar to those who exercise political power.

I share your view that a socialist economy possesses advantages which definitely counterbalance its disadvantages whenever the management lives up, at least to some extent, to adequate standards. No doubt, the day will come when all nations (as far as such nations still exist) will be grateful to Russia for having demonstrated, for the first time, by vigorous action the practical possibility of planned economy in spite of exceedingly great difficulties. I also believe that capitalism, or, we should say, the system of free enterprise, will prove unable to check unemployment, which will become increasingly chronic because of technological progress, and unable to maintain a healthy balance between production and the purchasing power of the people.

On the other hand we should not make the mistake of blaming capitalism for all existing social and political evils, and of assuming that the very establishment of socialism would be able to cure all the social and political ills of humanity. The danger of such a belief lies, first, in the fact that it encourages fanatical intolerance on the part of all the "faithful" by making a possible social method into a type of church which brands all those who do not belong to it as traitors or as nasty evil-doers. Once this stage has been reached, the ability to understand the convictions and actions of the "unfaithful" vanishes completely. You know, I am sure, from history how much unnecessary suffering such rigid beliefs have inflicted upon mankind.

Any government is in itself an evil in so far as it carries within it the tendency to deteriorate into tyranny. However, except for a very small number of anarchists, every one of us is convinced that civilized society cannot exist without a government. In a healthy nation there is a kind of dynamic balance between the will of the people and the government, which prevents its degeneration into tyranny. It is obvious that the danger of such deterioration is more acute in a country in which the government has authority not only over the armed forces but also over all the channels of education and information as well as over the economic existence of every single citizen. I say this merely to indicate that socialism as such cannot be considered the solution to all social problems but merely as a framework within which such a solution is possible.

What has surprised me most in your general attitude, expressed in your letter, is the following aspect: You are such passionate opponents of anarchy in the economic sphere, and yet equally passionate advocates of anarchy, e.g., unlimited sovereignty, in the sphere of international politics. The proposition to curtail the sovereignty of individual states appears to you in itself reprehensible, as a kind of violation of a natural right. In addition, you try to prove that behind the idea of curtailing sovereignty the United States is hiding her intention of economic domination and exploitation of the rest of the world without going to war. You attempt to justify this indictment by analyzing in your fashion the individual actions of this government since the end of the last war. You attempt to show that the Assembly of the United Nations is a mere puppet show controlled by the United States and hence the American capitalists.

Such arguments impress me as a kind of mythology; they are not convincing. They make obvious, however, the deep estrangement among the intellectuals of our two countries which is the result of a regrettable and artificial mutual isolation. If a free personal exchange of views should be made possible and should be encouraged, the intellectuals, possibly more than anyone else, could help to create an atmosphere of mutual understanding between the two nations and their problems. Such an atmosphere is a necessary prerequisite for the fruitful development of political cooperation. However, since for the time being we depend upon the cumbersome method of "open letters" I shall want to indicate briefly my reaction to your arguments.

Nobody would want to deny that the influence of the economic oligarchy upon all branches of our public life is very powerful. This influence, however, should not be overestimated. Franklin Delano Roosevelt was elected president in spite of desperate opposition by these very powerful groups and was re-elected three times; and this took place at a time when decisions of great consequence had to be made.

Concerning the policies of the American Government since the end of the war, I am neither willing, nor able, nor entitled to justify or explain them. It cannot be denied, however, that the suggestions of the American Government with regard to atomic weapons represented at least an attempt toward the creation of a supranational security organization. If they were not acceptable, they could at least have served as a basis of discussion for a real solution of the problems of international security. It is, indeed, the attitude of the Soviet Government, that was partly negative and partly dilatory, which has made it so difficult for well-meaning people in this country to use their political influence as they would have wanted, and to oppose the "war mongers." With regard to the influence of the United States upon the United Nations Assembly, I wish to say that, in my opinion, it stems not only from the economic and military power of the United States but also from the efforts of the United States and the United Nations to lead toward a genuine solution of the security problem.

Concerning the controversial veto power, I believe that the efforts to eliminate it or to make it ineffective have their primary cause less in specific intentions of the United States than in the manner in which the veto privilege has been abused.

Let me come now to your suggestion that the policy of the United States seeks to obtain economic domination and exploitation of other nations. It is a precarious undertaking to say anything reliable about aims and intentions. Let us rather examine the objective factors involved. The United States is fortunate in producing all the important industrial products and foods in her own country, in sufficient quantities. The country also possesses almost all important raw materials. Because of her tenacious belief in "free enterprise," she cannot succeed in keeping the purchasing power of the people in balance with the productive capacity of the country. For these very same reasons there is a constant danger that unemployment will reach threatening dimensions.

Because of these circumstances the United States is compelled to emphasize her export trade. Without it, she could not permanently keep her total productive machinery fully utilized. These conditions would not be harmful if the exports were balanced by imports of about the same value. Exploitation of foreign nations would then consist in the fact that the labor value of exports would considerably exceed that of imports. However, every effort is being made to avoid this, since almost every import would make a part of the productive machinery idle.

This is why foreign countries are not able to pay for the export commodities of the United States, payment which, in the long run, would indeed be possible only through imports by the latter. This explains why a large portion of all the gold has come to the United States. On the whole, this gold cannot be utilized except for the purchase of foreign commodities, which, because of the reasons already stated, is not practicable. There it lies, this gold, carefully protected against theft, a monument to governmental wisdom and to economic science! The reasons which I have just indicated make it difficult for me to take the alleged exploitation of the world by the United States very seriously.

However, the situation just described has a serious political facet. The United States, for the reasons indicated, is compelled to ship part of its production to foreign countries. These exports are financed through loans which the United States is granting foreign countries. It is, indeed, difficult to imagine how these loans will ever be repaid. For all practical purposes, therefore, these loans must be considered gifts which may be used as weapons in the arena of power politics. In view of the existing conditions and in view of the general characteristics of human beings, this, I frankly admit, represents a real danger. Is it not true, however, that we have stumbled into a state of international affairs which tends to make every invention of our minds and every material good into a weapon and, consequently, into a danger for mankind?

This question brings us to the most important matter, in comparison to which everything else appears insignificant indeed. We all know that power politics, sooner or later, necessarily leads to war, and that war, under present circumstances, would mean a mass destruction of human beings and material goods, the dimensions of which are much, much greater than anything that has ever before happened in history.

Is it really unavoidable that, because of our passions and our inherited customs, we should be condemned to annihilate each other so thoroughly that nothing would be left over which would deserve to be conserved? Is it not true that all the controversies and differences of opinion which we have touched upon in our strange exchange of letters are insignificant pettinesses compared to the danger in which we all find ourselves? Should we not do everything in our power to eliminate the danger which threatens all nations alike?

If we hold fast to the concept and practice of unlimited sovereignty of nations it only means that each country reserves the right for itself of pursuing its objectives through warlike means. Under the circumstances, every nation must be prepared for that possibility; this means it must try with all its might to be superior to anyone else. This objective will dominate more and more our entire public life and will poison our youth long before the catastrophe is itself actually upon us. We must not tolerate this, however, as long as we still retain a tiny bit of calm reasoning and human feelings.

This alone is on my mind in supporting the idea of "World Government," without any regard to what other people may have in mind when working for the same objective. I advocate world government because I am convinced that there is no other possible way of eliminating the most terrible danger in which man has ever found himself. The objective of avoiding total destruction must have priority over any other objective.

I am sure you are convinced that this letter is written with all the seriousness and honesty at my command; I trust you will accept it in the same spirit.

ON RECEIVING THE ONE WORLD AWARD

From an address at Carnegie Hall, April 27, 1948. Published in Out of My Later Years, *New York: Philosophical Library, 1950.*

I am greatly touched by the signal honor which you have wished to confer upon me. In the course of my long life I have received from my fellow-men far more recognition than I deserve, and I confess that my sense of shame has always outweighed my pleasure therein. But never, on any previous occasion, has the pain so far outweighed the pleasure as now. For all of us who are concerned for peace and the triumph of reason and justice must today be keenly aware how small an influence reason and honest good-will exert upon events in the political field. But however that may be, and whatever fate may have in store for us, yet we may rest assured that without the tireless efforts of those who are concerned with the welfare of humanity as a whole, the lot of mankind would be still worse than in fact it even now is.

In this time of decisions so heavy with fate, what we must say to our fellow-citizens seems above all to be this: where belief in the omnipotence of physical force gets the upper hand in political life, this force takes on a life of its own, and proves stronger than the men who think to use force as a tool. The proposed militarization of the nation not only immediately threatens us with war; it will also slowly but surely destroy the democratic spirit and the dignity of the individual in our land. The assertion that events abroad force us to arm is wrong, we must combat it with all our strength. Actually, our own rearmament, through the reaction of other nations to it, will bring about that very situation on which its advocates seek to base their proposals.

There is only *one* path to peace and security: the path of supranational organization. One-sided armament on a national basis only heightens the general uncertainty and confusion without being an effective protection.

A MESSAGE TO INTELLECTUALS

From the message to the Peace Congress of Intellectuals at Wroclaw, never delivered, but released to the press on August 29, 1948.

We meet today, as intellectuals and scholars of many nationalities, with a deep and historic responsibility placed upon us. We have every reason to be grateful to our French and Polish colleagues whose initiative has assembled us here for a momentous objective: to use the influence of wise men in promoting peace and security throughout the world. This is the age-old problem with which Plato, as one of the first, struggled so hard: to apply reason and prudence to the solution of man's problems instead of yielding to atavist instincts and passions.

By painful experience we have learned that rational thinking does not suffice to solve the problems of our social life. Penetrating research and keen scientific work have often had tragic implications for mankind, producing, on the one hand, inventions which liberated man from exhausting physical labor, making his life easier and richer; but on the other hand, introducing a grave restlessness into his life, making him a slave to his technological environment, and—most catastrophic of all—creating the means for his own mass destruction. This, indeed, is a tragedy of overwhelming poignancy!

However poignant that tragedy is, it is perhaps even more tragic that, while mankind has produced many scholars so extremely successful in the field of science and technology, we have been for a long time so inefficient in finding adequate solutions to the many political conflicts and economic tensions which beset us. No doubt, the antagonism of economic interests within and among nations is largely responsible to a great extent for the dangerous and threatening condition in the world today. Man has not succeeded in developing political and economic forms of organization which would guarantee the peaceful coexistence of the nations of the world. He has not succeeded in building the kind of system which would eliminate the possibility of war and banish forever the murderous instruments of mass destruction.

We scientists, whose tragic destination has been to help in making the methods of annihilation more gruesome and more effective, must consider it our solemn and transcendent duty to do all in our power in preventing these weapons from being used for the brutal purpose for which they were invented. What task could possibly be more important for us? What social aim could be closer to our hearts? That is why this Congress has such a vital mission. We are here to take counsel with each other. We must build spiritual and scientific bridges linking the nations of the world. We must overcome the horrible obstacles of national frontiers.

In the smaller entities of community life, man has made some progress toward breaking down anti-social sovereignties. This is true, for example, of life within cities and, to a certain degree, even of society within individual states. In such communities tradition and education have had a moderating influence and have brought about tolerable relations among the peoples living within those confines. But in relations among separate states complete anarchy still prevails. I do not believe that we have made any genuine advance in this area during the last few thousand years. All too frequently conflicts among nations are still being decided by brutal power, by war. The unlimited desire for ever greater power seeks to become active and aggressive wherever and whenever the physical possibility offers itself.

Throughout the ages, this state of anarchy in international affairs has inflicted indescribable suffering and destruction upon mankind; again and again it has depraved the development of men, their souls and their well-being. At times it has almost annihilated whole areas.

However, the desire of nations to be constantly prepared for warfare has, in addition, still other repercussions upon the lives of men. The power of every state over its citizens has grown steadily during the last few hundred years, no less in countries where the power of the state has been exercised wisely, than in those where it has been used for brutal tyranny. The function of the state to maintain peaceful and ordered relations among and between its citizens has become increasingly complicated and extensive largely because of the concentration and centralization of the modern industrial apparatus. In order to protect its citizens from attacks from without a modern state requires a formidable, expanding military establishment. In addition, the state considers it necessary to educate its citizens for the possibilities of war, an "education" not only corrupting to the soul and spirit of the young, but also adversely affecting the mentality of adults. No country can avoid this corruption. It pervades the citizenry even in countries which do not harbor outspoken aggressive tendencies. The state has thus become a modern idol whose suggestive power few men are able to escape.

Education for war, however, is a delusion. The technological developments of the last few years have created a completely new military situation. Horrible weapons have been invented, capable of destroying in a few seconds huge masses of human beings and tremendous areas of territory. Since science has not yet found protection from these weapons, the modern state is no longer in a position to prepare adequately for the safety of its citizens.

How, then, shall we be saved?

Mankind can only gain protection against the danger of unimaginable destruction and wanton annihilation if a supranational organization has alone the authority to produce or possess these weapons. It is unthinkable, however, that nations under existing conditions would hand over such authority to a supranational organization unless the organization would have the legal right and duty to solve all the conflicts which in the past have led to war. The functions of individual states would be to concentrate more or less upon internal affairs; in their relation with other states they would deal only with issues and problems which are in no way conducive to endangering international security.

Unfortunately, there are no indications that governments yet realize that the situation in which mankind finds itself makes the adoption of revolutionary measures a compelling necessity. Our situation is not comparable to anything in the past. It is impossible, therefore, to apply methods and measures which at an earlier age might have been sufficient. We must revolutionize our thinking, revolutionize our actions, and must have the courage to revolutionize relations among the nations of the world. Clichés of yesterday will no longer do today, and will, no doubt, be hopelessly out of date tomorrow. To bring this home to men all over the world is the most important and most fateful social function intellectuals have ever had to shoulder. Will they have enough courage to overcome their own national ties to the extent that is necessary to induce the peoples of the world to change their deep-rooted national traditions in a most radical fashion?

A tremendous effort is indispensable. If it fails now, the supranational organization will be built later, but then it will have to be built upon the ruins of a large part of the now existing world. Let us hope that the abolition of the existing international anarchy will not need to be bought by a self-inflicted world catastrophe the dimensions of which none of us can possibly imagine. The time is terribly short. We must act now if we are to act at all.

WHY SOCIALISM?

From Monthly Review, *New York, May, 1949.*

Is it advisable for one who is not an expert on economic and social issues to express views on the subject of socialism? I believe for a number of reasons that it is.

Let us first consider the question from the point of view of scientific knowledge. It might appear that there are no essential methodological differences between astronomy and economics: scientists in both fields attempt to discover laws of general acceptability for a circumscribed group of phenomena in order to make the interconnection of these phenomena as clearly understandable as possible. But in reality such methodological differences do exist. The discovery of general laws in the field of economics is made difficult by the circumstance that observed economic phenomena are often affected by many factors which are very hard to evaluate separately. In addition, the experience which has accumulated since the beginning of the so-called civilized period of human history has—as is well known—been largely influenced and limited by causes which are by no means exclusively economic in nature. For example, most of the major states of history owed their existence to conquest. The conquering peoples established themselves, legally and economically, as the privileged class of the conquered country. They seized for themselves a monopoly of the land ownership and appointed a priesthood from among their own ranks. The priests, in control of education, made the class division of society into a permanent institution and created a system of values by which the people were thenceforth, to a large extent unconsciously, guided in their social behavior.

But historic tradition is, so to speak, of yesterday; nowhere have we really overcome what Thorstein Veblen called "the predatory phase" of human development. The observable economic facts belong to that phase and even such laws as we can derive from them are not applicable to other phases. Since the real purpose of socialism is precisely to overcome and advance beyond the predatory phase of human development, economic science in its present state can throw little light on the socialist society of the future.

Second, socialism is directed toward a social-ethical end. Science, however, cannot create ends and, even less, instill them in human beings; science, at most, can supply the means by which to attain certain ends. But the ends themselves are conceived by personalities with lofty ethical ideals and—if these ends are not stillborn, but vital and vigorous—are adopted and carried forward by those many human beings who, half-unconsciously, determine the slow evolution of society.

For these reasons, we should be on our guard not to overestimate science and scientific methods when it is a question of human problems; and we should not assume that experts are the only ones who have a right to express themselves on questions affecting the organization of society.

Innumerable voices have been asserting for some time now that human society is passing through a crisis, that its stability has been gravely shattered. It is characteristic of such a situation that individuals feel indifferent or even hostile toward the group, small or large, to which they belong. In order to illustrate my meaning, let me record here a personal experience. I recently discussed with an intelligent and well-disposed man the threat of another war, which in my opinion would seriously endanger the existence of mankind, and I remarked that only a supranational organization would offer protection from that danger. Thereupon my visitor, very calmly and coolly, said to me: "Why are you so deeply opposed to the disappearance of the human race?"

I am sure that as little as a century ago no one would have so lightly made a statement of this kind. It is the statement of a man who has striven in vain to attain an equilibrium within himself and has more or less lost hope of succeeding. It is the expression of a painful solitude and isolation from which so many people are suffering in these days. What is the cause? Is there a way out?

It is easy to raise such questions, but difficult to answer them with any degree of assurance. I must try, however, as best I can, although I am very conscious of the fact that our feelings and strivings are often contradictory and obscure and that they cannot be expressed in easy and simple formulas.

Man is, at one and the same time, a solitary being and a social being. As a solitary being, he attempts to protect his own existence and that of those who are closest to him, to satisfy his personal desires, and to develop his innate abilities. As a social being, he seeks to gain the recognition and affection of his fellow human beings, to share in their pleasures, to comfort them in their sorrows, and to improve their conditions of life. Only the existence of these varied, frequently conflicting strivings accounts for the special character of a man, and their specific combination determines the extent to which an individual can achieve an inner equilibrium and can contribute to the well-being of society. It is quite possible that the relative strength of these two drives is, in the main, fixed by inheritance. But the personality that finally emerges is largely formed by the environment in which a man happens to find himself during his development, by the structure of the society in which he grows up, by the tradition of that society, and by its appraisal of particular types of behavior. The abstract concept "society" means to the individual human being the sum total of his direct and indirect relations to his contemporaries and to all the people of earlier generations. The individual is able to think, feel, strive, and work by himself; but he depends so much upon society—in his physical, intellectual, and emotional existence—that it is impossible to think of him, or to understand him, outside the framework of society. It is "society" which provides man with food, clothing, a home, the tools of work, language, the forms of thought, and most of the content of thought; his life is made possible through the labor and the accomplishments of the many millions past and present who are all hidden behind the small word "society."

It is evident, therefore, that the dependence of the individual upon society is a fact of nature which cannot be abolished—just as in the case of ants and bees. However, while the whole life process of ants and bees is fixed down to the smallest detail by rigid, hereditary instincts, the social pattern and interrelationships of human beings are very variable and susceptible to change. Memory, the capacity to make new combinations, the gift of oral communication have made possible developments among human beings which are not dictated by biological necessities. Such developments manifest themselves in traditions, institutions, and organizations; in literature; in scientific and engineering accomplishments; in works of art. This explains how it happens that, in a certain sense, man can influence his life through his own conduct, and that in this process conscious thinking and wanting can play a part.

Man acquires at birth, through heredity, a biological constitution which we must consider fixed and unalterable, including the natural urges which are characteristic of the human species. In addition, during his lifetime, he acquires a cultural constitution which he adopts from society through communication and through many other types of influences. It is this cultural constitution which, with the passage of time, is subject to change and which determines to a very large extent the relationship between the individual and society.

Modern anthropology has taught us, through comparative investigation of so-called primitive cultures, that the social behavior of human beings may differ greatly, depending upon prevailing cultural patterns and the types of organization which predominate in society. It is on this that those who are striving to improve the lot of man may ground their hopes: human beings are *not* condemned, because of their biological constitution, to annihilate each other or to be at the mercy of a cruel, self-inflicted fate.

If we ask ourselves how the structure of society and the cultural attitude of man should be changed in order to make human life as satisfying as possible, we should constantly be conscious of the fact that there are certain conditions which we are unable to modify. As mentioned before, the biological nature of man is, for all practical purposes, not subject to change. Furthermore, technological and demographic developments of the last few centuries have created conditions which are here to stay. In relatively densely settled populations with the goods which are indispensable to their continued existence, an extreme division of labor and a highly centralized productive apparatus are absolutely necessary. The time—which, looking back, seems so idyllic—is gone forever when individuals or relatively small groups could be completely self-sufficient. It is only a slight exaggeration to say that mankind constitutes even now a planetary community of production and consumption.

I have now reached the point where I may indicate briefly what to me constitutes the essence of the crisis of our time. It concerns the relationship of the individual to society. The individual has become more conscious than ever of his dependence upon society. But he does not experience this dependence as a positive asset, as an organic tie, as a protective force, but rather as a threat to his natural rights, or even to his economic existence. Moreover, his position in society is such that the egotistical drives of his make-up are constantly being accentuated, while his social drives, which are by nature weaker, progressively deteriorate. All human beings, whatever their position in society, are suffering from this process of deterioration. Unknowingly prisoners of their own egotism, they feel insecure, lonely, and deprived of the naïve. simple, and unsophisticated enjoyment of life. Man can find meaning in life, short and perilous as it is, only through devoting himself to society.

The economic anarchy of capitalist society as it exists today is, in my opinion, the real source of the evil. We see before us a huge community of producers the members of which are unceasingly striving to deprive each other of the fruits of their collective labor—not by force, but on the whole in faithful compliance with legally established rules. In this respect, it is important to realize that the means of production—that is to say, the entire productive capacity that is needed for producing consumer goods as well as additional capital goods —may legally be, and for the most part are, the private property of individuals.

For the sake of simplicity, in the discussion that follows I shall call "workers" all those who do not share in the ownership of the means of production—although this does not quite correspond to the customary use of the term. The owner of the means of production is in a position to purchase the labor power of the worker. By using the means of production, the worker produces new goods which become the property of the capitalist. The essential point about this process is the relation between what the worker produces and what he is paid, both measured in terms of real value. In so far as the labor contract is "free," what the worker receives is determined not by the real value of the goods he produces, but by his minimum needs and by the capitalists' requirements for labor power in relation to the number of workers competing for jobs. It is important to understand that even in theory the payment of the worker is not determined by the value of his product.

Private capital tends to become concentrated in few hands, partly because of competition among the capitalists, and partly because technological development and the increasing division of labor encourage the formation of larger units of production at the expense of the smaller ones. The result of these developments is an oligarchy of private capital the enormous power of which cannot be effectively checked even by a democratically organized political society. This is true since the members of legislative bodies are selected by political parties, largely financed or otherwise influenced by private capitalists who, for all practical purposes, separate the electorate from the legislature. The consequence is that the representatives of the people do not in fact sufficiently protect the interests of the underprivileged sections of the population. Moreover, under existing conditions, private capitalists inevitably control, directly or indirectly, the main sources of information (press, radio, education). It is thus extremely difficult, and indeed in most cases quite impossible, for the individual citizen to come to objective conclusions and to make intelligent use of his political rights.

The situation prevailing in an economy based on the private ownership of capital is thus characterized by two main principles: first, means of production (capital) are privately owned and the owners dispose of them as they see fit; second, the labor contract is free. Of course, there is no such thing as a *pure* capitalist society in this sense. In particular, it should be noted that the workers, through long and bitter political struggles, have succeeded in securing a somewhat improved form of the "free labor contract" for certain categories of workers. But taken as a whole, the present-day economy does not differ much from "pure" capitalism.

Production is carried on for profit, not for use. There is no provision that all those able and willing to work will always be in a position to find employment; an "army of unemployed" almost always exists. The worker is constantly in fear of losing his job. Since unemployed and poorly paid workers do not provide a profitable market, the production of consumers' goods is restricted, and great hardship is the consequence. Technological progress frequently results in more unemployment rather than in an easing of the burden of work for all. The profit motive, in conjunction with competition among capitalists, is responsible for an instability in the accumulation and utilization of capital which leads to increasingly severe depressions. Unlimited competition leads to a huge waste of labor, and to that crippling of the social consciousness of individuals which I mentioned before.

This crippling of individuals I consider the worst evil of capitalism. Our whole educational system suffers from this evil. An exaggerated competitive attitude is inculcated into the student, who is trained to worship acquisitive success as a preparation for his future career.

I am convinced there is only *one* way to eliminate these grave evils, namely through the establishment of a socialist economy, accompanied by an educational system which would be oriented toward social goals. In such an economy, the means of production are owned by society itself and are utilized in a planned fashion. A planned economy, which adjusts production to the needs of the community, would distribute the work to be done among all those able to work and would guarantee a livelihood to every man, woman, and child. The education of the individual, in addition to promoting his own innate abilities, would attempt to develop in him a sense of responsibility for his fellow-men in place of the glorification of power and success in our present society.

Nevertheless, it is necessary to remember that a planned economy is not yet socialism. A planned economy as such may be accompanied by the complete enslavement of the individual. The achievement of socialism requires the solution of some extremely difficult socio-political problems: how is it possible, in view of the far-reaching centralization of political and economic power, to prevent bureaucracy from becoming all-powerful and overweening? How can the rights of the individual be protected and therewith a democratic counterweight to the power of bureaucracy be assured?

NATIONAL SECURITY

Contribution to Mrs. Eleanor Roosevelt's television program concerning the implications of the H-bomb, February 13, 1950.

I am grateful to you, Mrs. Roosevelt, for the opportunity to express my conviction in this most important political question.

The idea of achieving security through national armament is, at the present state of military technique, a disastrous illusion. On the part of the U.S.A. this illusion has been particularly fostered by the fact that this country succeeded first in producing an atomic bomb. The belief seemed to prevail that in the end it would be possible to achieve decisive military superiority. In this way, any potential opponent would be intimidated, and security, so ardently desired by all of us, brought to us and all of humanity. The maxim which we have been following during these last five years has been, in short: security through superior military power, whatever the cost.

This mechanistic, technical-military psychological attitude has had its inevitable consequences. Every single act in foreign policy is governed exclusively by one viewpoint: how do we have to act in order to achieve utmost superiority over the opponent in case of war? Establishing military bases at all possible strategically important points on the globe. Arming and economic strengthening of potential allies. Within the country: concentration of tremendous financial power in the hands of the military; militarization of the youth; close supervision of the loyalty of the citizens, in particular, of the civil servants, by a police force growing more conspicuous every day. Intimidation of people of independent political thinking. Subtle indoctrination of the public by radio, press, and schools. Growing restriction of the range of public information under the pressure of military secrecy.

The armament race between the U.S.A. and the U.S.S.R.,

originally supposed to be a preventive measure, assumes hysterical character. On both sides, the means to mass destruction are perfected with feverish haste—behind the respective walls of secrecy. The hydrogen bomb appears on the public horizon as a probably attainable goal. Its accelerated development has been solemnly proclaimed by the President. If it is successful, radioactive poisoning of the atmosphere and hence annihilation of any life on earth has been brought within the range of technical possibilities. The ghostlike character of this development lies in its apparently compulsory trend. Every step appears as the unavoidable consequence of the preceding one. In the end, there beckons more and more clearly general annihilation.

Is there any way out of this impasse created by man himself? All of us, and particularly those who are responsible for the attitude of the U.S.A. and the U.S.S.R., should realize that we may have vanquished an external enemy, but have been incapable of getting rid of the mentality created by the war. It is impossible to achieve peace as long as every single action is taken with a possible future conflict in view. The leading point of view of all political action should therefore be: what can we do to bring about a peaceful coexistence and even loyal cooperation of the nations? The first problem is to do away with mutual fear and distrust. Solemn renunciation of violence (not only with respect to means of mass destruction) is undoubtedly necessary. Such renunciation, however, can be effective only if at the same time a supranational judicial and executive body is set up empowered to decide questions of immediate concern to the security of the nations. Even a *declaration* of the nations to collaborate loyally in the realization of such a "restricted world government" would considerably reduce the imminent danger of war.

In the last analysis, every kind of peaceful cooperation among men is primarily based on mutual trust and only secondly on institutions such as courts of justice and police. This holds for nations as well as for individuals. And the basis of trust is loyal give and take.

What about international control? Well, it may be of secondary use as a police measure. But it may be wise not to overestimate its importance. The times of Prohibition come to mind
and give one pause.

THE PURSUIT OF PEACE

*U.N. radio interview, June 16, 1950, recorded in the study
of Einstein's Princeton, N. J., home.*

Q: Is it an exaggeration to say that the fate of the world is
hanging in the balance?

A: No exaggeration. The fate of humanity is always in the
balance . . . but more truly now than at any known time.

Q: How can we awaken all the peoples to the seriousness of
the moment?

A: I believe this *can* be answered. A remedy can't be found
in preparing for the event of war, but in starting from the conviction that security from military disaster can be realized only
by patient negotiation and through creation of a legal basis for
the solution of international problems, supported by a sufficiently strong executive agency—in short, a kind of world
government.

Q: Is the current atomic armaments race leading to another
world war or—as some people maintain—a way to prevent war?

A: Competitive armament is not a way to prevent war. Every
step in this direction brings us nearer to catastrophe. The
armaments race is the *worst* method to prevent open conflict.
On the contrary, real peace cannot be reached without systematic disarmament on a supranational scale. I repeat, armament is no protection against war, but leads inevitably *to* war.

Q: Is it possible to prepare for war and a world community
at the same time?

A: Striving for peace and preparing for war are incompatible with each other, and in our time more so than ever.

Q: Can we prevent war?

A: There is a very simple answer. If we have the courage to decide ourselves for peace, we will *have* peace.

Q: How?

A: By the firm *will* to reach agreement. This is axiomatic. We are not engaged in a play but in a condition of utmost danger to existence. If you are not firmly decided to resolve things in a peaceful way, you will never come to a peaceful solution.

Q: What is your estimate of the future effect of atomic energy on our civilization in the next ten or twenty years?

A: Not relevant now. The technical possibilities we now have already are satisfactory enough . . . if the right use would be made of them.

Q: What is your opinion of the profound changes in our living predicted by some scientists . . . for example, the possibility of our need to work only two hours a day?

A: We are always the same people. There are not really profound changes. It is not so important if we work five hours or two. Our problem is social and economic, at the international level.

Q: What would you suggest doing with the present supply of atom bombs already stockpiled?

A: Give it to a supranational organization. During the interval before solid peace one must have some protecting power. One-sided disarmament is not possible; this is out of the question. Arms must be entrusted only to an international authority. There is no other possibility . . . systematic disarmament connected with supranational government. One must not look too technically on the problem of security. The *will* to peace and the readiness to accept every step needed for this goal are most important.

Q: What can a private individual *do* about war or peace?

A: Individuals can cause anyone who tries to be elected (for Congress, etc.) to give clear promise to work for international order and restriction of national sovereignty in favor of that order. Everybody is involved in forming public opinion . . . and he must really understand what is needed . . . and he must have the courage to speak out.

Q: United Nations Radio is broadcasting to all the corners of the earth, in twenty-seven languages. Since this is a moment of great danger, what word would you have us broadcast to the peoples of the world?

A: Taken on the whole, I would believe that Gandhi's views were the most enlightened of all the political men in our time. We should strive to do things in his spirit . . . not to use violence in fighting for our cause, but by non-participation in what we believe is evil.

"CULTURE MUST BE ONE OF THE FOUNDATIONS FOR WORLD UNDERSTANDING"

From Unesco Courier, *December, 1951.*

In order to grasp the full significance of the Universal Declaration of Human Rights, it is well to be fully aware of the world situation that gave birth to the United Nations and to Unesco. The devastation wrought by the wars of the last half century had brought home the fact to everybody that, with the present-day level of technical achievement, the security of nations could be based only on supranational institutions and rules of conduct. It is understood that, in the long run, an all-destroying conflict can be avoided only by the setting up of a world federation of nations.

So—as a modest beginning of international order—the United Nations was founded. This organization, however, is but a meeting ground for delegates of national governments and not for the peoples' representatives acting independently on the basis of their own personal convictions. Furthermore, U.N. decisions do not have binding force on any national government; nor do any concrete means exist by which the decisions can be enforced.

The effectiveness of the United Nations is still further reduced by the fact that membership has been refused to certain nations, whose exclusion seriously affects the supra character of the organization. Yet, in itself, the fact that international problems are brought up and discussed in the broad light of day favors the peaceful solution of conflicts. The existence of a supranational platform of discussion is apt to accustom the peoples gradually to the idea that national interests must be safeguarded by negotiation and not by brute force.

This psychological or educational effect I regard as the United Nations' most valuable feature. A world federation presupposes a new kind of loyalty on the part of man, a sense of responsibility that does not stop short at the national boundaries. To be truly effective, such loyalty must embrace more than purely political issues. Understanding among different cultural groups, mutual economic and cultural aid are the necessary additions.

Only by such endeavor will the feeling of confidence be established that was lost owing to the psychological effect of the wars and sapped by the narrow philosophy of militarism and power politics. No effective institution for the collective security of nations is possible without understanding and a measure of reciprocal confidence.

To the U.N. was added Unesco, the agency whose function it is to pursue these cultural tasks. It has in a greater measure than U.N. been able to avoid the paralyzing influence of power politics.

Realizing that healthy international relations can be created only among populations made up of individuals who themselves are healthy and enjoy a measure of independence, the United Nations elaborated a Universal Declaration of Human Rights, which was adopted by the U.N. General Assembly on December 10, 1948.

The Declaration establishes a number of universally comprehensible standards that are designed to protect the individual, to prevent his being exploited economically, and to safeguard his development and the free pursuit of his activities within the social framework.

To spread these standards among all U.N. Member States is rightly regarded and aimed at as an important objective. Unesco has accordingly instituted this third celebration for the purpose of drawing attention far and wide to these fundamental aspirations as a basis on which to restore the political health of the peoples.

It was scarcely to be avoided that the Declaration should take the form of a legalistic document, which in its rigidity may lead to endless discussion. It is impossible for such a text to take the great diversity of conditions of life in the different countries fully into account; in addition, it is unavoidable that such a text admits various interpretations of detail. The general tendency of the Declaration, however, is unmistakable and provides a suitable, generally acceptable basis for judgment and action.

To give formal recognition to standards and to adopt them as the guiding lines of action in the teeth of all the adversities of a changing situation are two very different things—as the impartial observer may see particularly in the history of religious institutions. Then and only then will the Declaration exert effective influence, when the United Nations itself shows by its decisions and actions that it does embody, *de facto*, the spirit of this, its own Declaration.

ON THE ABOLITION OF THE THREAT OF WAR

Written September 20, 1952. Published in Japanese maga-
zine, Kaizo *(Tokyo), Autumn, 1952.*

My part in producing the atomic bomb consisted in a single act: I signed a letter to President Roosevelt, pressing the need for experiments on a large scale in order to explore the possibilities for the production of an atomic bomb.

I was fully aware of the terrible danger to mankind in case this attempt succeeded. But the likelihood that the Germans were working on the same problem with a chance of succeeding forced me to this step. I could do nothing else although I have always been a convinced pacifist. To my mind, to kill in war is not a whit better than to commit ordinary murder.

As long, however, as the nations are not resolved to abolish war through common actions and to solve their conflicts and protect their interests by peaceful decisions on a legal basis, they feel compelled to prepare for war. . They feel obliged to prepare all possible means, even the most detestable ones, so as not to be left behind in the general armament race. This road necessarily leads to war, a war which under the present conditions means universal destruction.

Under these circumstances the fight against *means* has no chance of success. Only the radical abolition of wars and of the threat of war can help. This is what one has to work for. One has to be resolved not to let himself be forced to actions that run counter to this goal. This is a severe demand on an individual who is conscious of his dependence on society. But it is not an impossible demand.

Gandhi, the greatest political genius of our time, has pointed the way. He has shown of what sacrifices people are capable once they have found the right way. His work for the liberation of India is a living testimony to the fact that a will governed by firm conviction is stronger than a seemingly invincible material power.

SYMPTOMS OF CULTURAL DECAY

Bulletin of Atomic Scientists, *Vol. VIII, No. 7, October, 1952.*

The free, unhampered exchange of ideas and scientific conclusions is necessary for the sound development of science, as it is in all spheres of cultural life. In my opinion, there can be no doubt that the intervention of political authorities of this country in the free exchange of knowledge between individuals has already had significantly damaging effects. First of all, the damage is to be seen in the field of scientific work proper, and, after a while, it will become evident in technology and industrial production.

The intrusion of the political authorities into the scientific life of our country is especially evident in the obstruction of the

travels of American scientists and scholars abroad and of for-
eign·scientists seeking to come to this country. Such petty behavior on the part of a powerful country is only a peripheral
symptom of an ailment which has deeper roots.

Interference with the freedom of the oral and written communication of scientific results, the widespread attitude of political distrust which is supported by an immense police organization, the timidity and the anxiety of individuals to avoid
everything which might cause suspicion and which could
threaten their economic position—all these are only symptoms,
even though they reveal more clearly the threatening character
of the illness.

The real ailment, however, seems to me to lie in the attitude
which was created by the World War and which dominates all
our actions; namely, the belief that we must in peacetime so
organize our whole life and work that in the event of war we
would be sure of victory. This attitude gives rise to the belief
that one's freedom and indeed one's existence are threatened
by powerful enemies.

This attitude explains all of the unpleasant facts which we
have designated above as symptoms. It must, if it does not
rectify itself, lead to war and to very far-reaching destruction.
It finds its expression in the budget of the United States.

Only if we overcome this obsession can we really turn our
attention in a reasonable way to the real political problem,
which is, "How can we contribute to make the life of man on
this diminishing earth more secure and more tolerable?"

It will be impossible to cure ourselves of the symptoms we
have mentioned and many others if we do not overcome the
deeper ailment which is affecting us.

Part III

ON THE JEWISH PEOPLE

A LETTER TO PROFESSOR DR. HELLPACH, MINISTER OF STATE

Written in response to an article by Professor Hellpach which appeared in the Vossische Zeitung *in 1929. Published in* Mein Weltbild, *Amsterdam: Querido Verlag, 1934.*

DEAR MR. HELLPACH:

I have read your article on Zionism and the Zurich Congress and feel, as a strong devotee of the Zionist idea, that I must answer you, even if only shortly.

The Jews are a community bound together by ties of blood and tradition, and not of religion only: the attitude of the rest of the world toward them is sufficient proof of this. When I came to Germany fifteen years ago I discovered for the first time that I was a Jew, and I owe this discovery more to Gentiles than Jews.

The tragedy of the Jews is that they are people of a definite historical type, who lack the support of a community to keep them together. The result is a want of solid foundations in the individual which amounts in its extremer forms to moral instability. I realized that salvation was only possible for the race if every Jew in the world should become attached to a living society to which he as an individual might rejoice to belong and which might enable him to bear the hatred and the humiliations that he has to put up with from the rest of the world.

I saw worthy Jews basely caricatured, and the sight made my heart bleed. I saw how schools, comic papers, and innumerable other forces of the Gentile majority undermined the confidence even of the best of my fellow-Jews, and felt that this could not be allowed to continue.

171

Then I realized that only a common enterprise dear to the heart of Jews all over the world could restore this people to health. It was a great achievement of Herzl's to have realized and proclaimed at the top of his voice that, the traditional attitude of the Jews being what it was, the establishment of a national home or, more accurately, a center in Palestine, was a suitable object on which to concentrate our efforts.

All this you call nationalism, and there is something in the accusation. But a communal purpose without which we can neither live nor die in this hostile world can always be called by that ugly name. In any case it is a nationalism whose aim is not power but dignity and health. If we did not have to live among intolerant, narrow-minded, and violent people, I should be the first to throw over all nationalism in favor of universal humanity.

The objection that we Jews cannot be proper citizens of the German state, for example, if we want to be a "nation," is based on a misunderstanding of the nature of the state which springs from the intolerance of national majorities. Against that intolerance we shall never be safe, whether we call ourselves a people (or nation) or not.

I have put all this with brutal frankness for the sake of brevity, but I know from your writings that you are a man who stands to the sense, not the form.

LETTER TO AN ARAB

Mein Weltbild, *Amsterdam: Querido Verlag, 1934.*

March 15, 1930

SIR:

Your letter has given me great pleasure. It shows me that there is good will available on your side, too, for solving the present difficulties in a manner worthy of both our nations. I believe that these difficulties are more psychological than real, and that they can be got over if both sides bring honesty and good will to the task.

What makes the present position so bad is the fact that Jews and Arabs confront each other as opponents before the mandatory power. This state of affairs is unworthy of both nations and can only be altered by our finding a *via media* on which both sides agree.

I will now tell you how I think that the present difficulties might be remedied; at the same time I must add that this is only my personal opinion, which I have discussed with nobody. I am writing this letter in German because I am not capable of writing it in English myself and because I want to bear the entire responsibility for it myself. You will, I am sure, be able to get some Jewish friend of conciliation to translate it.

A Privy Council is to be formed to which the Jews and Arabs shall each send four representatives, who must be independent of all political parties:—

Each group to be composed as follows:—

A doctor, elected by the Medical Association.

A lawyer, elected by the lawyers.

A working men's representative, elected by the trade unions.

An ecclesiastic, elected by the ecclesiastics.

These eight people are to meet once a week. They undertake not to espouse the sectional interests of their profession or nation but conscientiously and to the best of their power to aim at the welfare of the whole population of the country. Their deliberations shall be secret and they are strictly forbidden to give any information about them, even in private. When a decision has been reached on any subject in which not less than three members on each side concur, it may be published, but only in the name of the whole Council. If a member dissents he may retire from the Council, but he is not thereby released from the obligation to secrecy. If one of the elective bodies above specified is dissatisfied with a resolution of the Council, it may replace its representative by another.

Even if this "Privy Council" has no definite powers, it may nevertheless bring about the gradual composition of differences, and secure a united representation of the common interests of

the country before the mandatory power, clear of the dust of
ephemeral politics.

THE JEWISH COMMUNITY

*A speech delivered at the Savoy Hotel, London, October
29, 1930. Published in* Mein Weltbild, *Amsterdam: Querido Verlag, 1934.*

LADIES AND GENTLEMEN:

It is no easy matter for me to overcome my natural inclination
to a life of quiet contemplation. But I could not remain deaf
to the appeal of the ORT and OZE societies *; for in responding to it I am responding, as it were, to the appeal of our sorely
oppressed Jewish nation.

The position of our scattered Jewish community is a moral
barometer for the political world. For what surer index of political morality and respect for justice can there be than the
attitude of the nations toward a defenseless minority, whose
peculiarity lies in their preservation of an ancient cultural tradition?

This barometer is low at the present moment, as we are
painfully aware from the way we are treated. But it is this
very lowness that confirms me in the conviction that it is our
duty to preserve and consolidate our community. Embedded
in the tradition of the Jewish people there is a love of justice
and reason which must continue to work for the good of all
nations now and in the future. In modern times this tradition
has produced Spinoza and Karl Marx.

Those who would preserve the spirit must also look after
the body to which it is attached. The OZE society literally looks
after the bodies of our people. In Eastern Europe it is working
day and night to help our people there, on whom the economic
depression has fallen particularly heavily, to keep body and soul
together; while the ORT society is trying to get rid of a severe

---

* Jewish charitable associations.

social and economic handicap under which the Jews have labored since the Middle Ages. Because we were then excluded from all directly productive occupations, we were forced into the purely commercial ones. The only way of really helping the Jew in eastern countries is to give him access to new fields of activity, for which he is struggling all over the world. This is the grave problem which the ORT society is successfully tackling.

It is to you English fellow-Jews that we now appeal to help us in this great enterprise which splendid men have set on foot. The last few years, nay, the last few days have brought us a disappointment which must have touched you particularly. Do not gird at fate but rather look on these events as a reason for remaining true to the cause of the Jewish commonwealth. I am convinced that in doing so we shall also indirectly be promoting those general human ends which we must always recognize as the highest.

Remember that difficulties and obstacles are a valuable source of health and strength to any society. We should not have survived for thousands of years as a community if our bed had been of roses; of that I am quite sure.

But we have a still fairer consolation. Our friends are not exactly numerous, but among them are men of noble spirit endowed with a strong sense of justice, who have devoted their lives to uplifting human society and liberating the individual from degrading oppression.

*   *   *

To you all I say that the existence and destiny of our people depends less on external factors than on ourselves. It is our duty to remain faithful to the moral traditions which have enabled us to survive for thousands of years despite the heavy storms that have broken over our heads. In the service of life sacrifice becomes grace.

ADDRESSES ON RECONSTRUCTION IN PALESTINE

*From 1920 on, observing the spread of anti-Semitism in
Germany after World War I, Einstein, who up to that time
had expressed little interest in religious matters, became
a strong supporter of the Zionist movement. In 1921 he
came to New York, with Professor Chaim Weizmann, later
to become the first president of the State of Israel, to raise
funds for the Jewish National Fund and the Hebrew University in Jerusalem (founded in 1918). The first three
talks below were delivered, however, during his third visit
to the United States in 1931–32. (His second American
visit had occurred in 1930.) The fourth talk was made
many years earlier upon his return from America to Berlin
in 1921, while the fifth, though more recent, nevertheless
pre-dated his settling in Princeton (1933). All were published in* Mein Weltbild, *Amsterdam: Querido Verlag,
1934.*

I.

Ten years ago, when I first had the pleasure of addressing you
in behalf of the Zionist cause, almost all our hopes were still
fixed on the future. Today we can look back on these ten
years with joy; for in that time the united energies of the Jewish people have accomplished a piece of splendidly successful,
constructive work in Palestine, which certainly exceeds anything
that we dared then to hope for.

We have also successfully stood the severe test to which the
events of the last few years have subjected us. Ceaseless work,
supported by a noble purpose, is leading slowly but surely to
success. The latest pronouncements of the British government
indicate a return to a juster judgment of our case; this we recognize with gratitude.

But we must never forget what this crisis has taught us—
namely, that the establishment of satisfactory relations between
the Jews and the Arabs is not England's affair but ours. We—

that is to say, the Arabs and ourselves—have got to agree on the main outlines of an advantageous partnership which shall satisfy the needs of both nations. A just solution of this problem and one worthy of both nations is an end no less important and no less worthy of our efforts than the promotion of the work of construction itself. Remember that Switzerland represents a higher stage of political development than any national state, precisely because of the greater political problems which had to be solved before a stable community could be built up out of groups of different nationality.

Much remains to be done, but one at least of Herzl's aims has already been realized: the Palestine job has given the Jewish people an astonishing degree of solidarity and the optimism without which no organism can lead a healthy life.

Anything we may do for the common purpose is done not merely for our brothers in Palestine but for the well-being and honor of the whole Jewish people.

II.

We are assembled today for the purpose of recalling to mind our age-old community, its destiny and its problems. It is a community of moral tradition, which has always shown its strength and vitality in times of stress. In all ages it has produced men who embodied the conscience of the western world, defenders of human dignity and justice.

So long as we ourselves care about this community it will continue to exist to the benefit of mankind, in spite of the fact that it possesses no self-contained organization. A decade or two ago a group of far-sighted men, among whom the unforgettable Herzl stood out above the rest, came to the conclusion that we needed a spiritual center in order to preserve our sense of solidarity in difficult times. Thus arose the idea of Zionism and the work of settlement in Palestine, the successful realization of which we have been permitted to witness, at least in its highly promising beginnings.

I have had the privilege of seeing, to my great joy and satisfaction, how much this achievement has contributed to the convalescence of the Jewish people; for the Jews are exposed, as a minority among the nations, not merely to external dangers but also to internal ones of a psychological nature.

The crisis which the work of construction has had to face in the last few years has lain heavy upon us and is not yet completely surmounted. But the most recent reports show that the world, and especially the British government, is disposed to recognize the great things which lie behind our struggle for the Zionist ideal. Let us at this moment remember with gratitude our leader Weizmann, whose zeal and circumspection helped the good cause to success.

The difficulties we have been through have also brought some good in their train. They have shown us once more how strong is the bond which unites the Jews of all countries in a common destiny. The crisis has also purified our attitude to the question of Palestine, purged it of the dross of nationalism. It has been clearly proclaimed that we are not seeking to create a political society, but that our aim is, in accordance with the old tradition of Jewry, a cultural one in the widest sense of the world. That being so, it is for us to solve the problem of living side by side with our brother the Arab in an open, generous, and worthy manner. We have here an opportunity of showing what we have learned in the thousands of years of our martyrdom. If we choose the right path, we shall succeed and give the rest of the world a fine example.

Whatever we do for Palestine, we do it for the honor and well-being of the whole Jewish people.

III.

I am delighted to have the opportunity of addressing a few words to the youth of this country which is faithful to the common aims of Jewry. Do not be discouraged by the difficulties which confront us in Palestine. Such things serve to test the will to live of our community.

Certain proceedings and pronouncements of the English administration have been justly criticized. We must not, however, let the matter rest at that, but draw what lesson we can from the experience.

We need to pay great attention to our relations with the Arabs. By cultivating these carefully we shall be able in future to prevent things from becoming so dangerously strained that people can take advantage of them to provoke acts of hostility. This goal is perfectly within our reach, because our work of construction has been, and must continue to be, carried out in such a manner as to serve the real interests of the Arab population also.

In this way we shall be able to avoid getting ourselves quite so often into the position, disagreeable for Jews and Arabs alike, of having to call in the mandatory power as arbitrator. We shall thereby be following not merely the dictates of Providence but also our traditions, which alone give the Jewish community meaning and stability. For our community is not, and must never become, a political one; this is the only permanent source whence it can draw new strength and the only ground on which its existence can be justified.

IV.

For the last two thousand years the common property of the Jewish people has consisted entirely of its past. Scattered over the wide world, our nation possessed nothing in common except its carefully guarded tradition. Individual Jews no doubt produced great work, but it seemed as if the Jewish people as a whole had not the strength left for great collective achievements.

Now all that is changed. History has set us a great and noble task in the shape of active cooperation in the building up of Palestine. Eminent members of our race are already at work with all their might on the realization of this aim. The opportunity is presented to us of setting up centers of civilization which the whole Jewish people can regard as its work. We nurse the hope of erecting in Palestine a home of our own national culture which shall help to awaken the Near East to new economic and spiritual life.

The object which the leaders of Zionism have before their

eyes is not a political but a social and cultural one. The community in Palestine must approach the social ideal of our forefathers as it is laid down in the Bible, and at the same time become a seat of modern intellectual life, a spiritual center for the Jews of the whole world. In accordance with this notion, the establishment of a Jewish university in Jerusalem constitutes one of the most important aims of the Zionist organization.

During the last few months I have been to America in order to help raise the material basis for this University there. The success of this enterprise was a natural one. Thanks to the untiring energy and splendid self-sacrificing spirit of the Jewish doctors in America we have succeeded in collecting enough money for the creation of a Medical Faculty, and the preliminary work is being started at once. After this success I have no doubt that the material basis for the other faculties will soon be forthcoming. The Medical Faculty is first of all to be developed as a research institute and to concentrate on making the country healthy, a most important item in the work of development. Teaching on a large scale will only become important later on. As a number of highly competent scientific workers have already signified their readiness to take up appointments at the University, the establishment of a Medical Faculty seems to be placed beyond all doubt. I may add that a special fund for the University, entirely distinct from the general fund for the development of the country, has been opened. For the latter, considerable sums have been collected during these months in America, thanks to the indefatigable labors of Professor Weizmann and other Zionist leaders, chiefly through the self-sacrificing spirit of the middle classes. I conclude with a warm appeal to the Jews in Germany to contribute all they can, in spite of the present economic difficulties, for the building up of the Jewish home in Palestine. This is not a matter of charity but an enterprise which concerns all Jews and the success of which promises to be a source of the highest satisfaction to all.

V.

For us Jews, Palestine is not just a charitable or colonial enterprise, but a problem of central importance for the Jewish people. Palestine is not primarily a place of refuge for the Jews of Eastern Europe but the embodiment of the re-awakening corporate spirit of the whole Jewish nation. Is it the right moment for this corporate sense to be awakened and strengthened? This is a question to which I feel compelled, not merely by my spontaneous feelings but on rational grounds, to return an unqualified "yes."

Let us just cast our eyes over the history of the Jews in Germany during the past hundred years. A century ago our forefathers, with few exceptions, lived in the ghetto. They were poor, without political rights, separated from the Gentiles by a barrier of religious traditions, habits of life, and legal restrictions; their intellectual development was restricted to their own literature, and they had remained almost unaffected by the mighty advance of the European intellect which dates from the Renaissance. And yet these obscure, humble people had one great advantage over us: each of them belonged in every fiber of his being to a community in which he was completely absorbed, in which he felt himself a fully privileged member, and which demanded nothing of him that was contrary to his natural habit of thought. Our forefathers in those days were pretty poor specimens intellectually and physically, but socially speaking they enjoyed an enviable spiritual equilibrium.

Then came emancipation, which suddenly opened up undreamed-of possibilities to the individual. Some few rapidly made a position for themselves in the higher walks of business and social life. They greedily lapped up the splendid triumphs which the art and science of the western world had achieved. They joined in the process with burning enthusiasm, themselves making contributions of lasting value. At the same time they imitated the external forms of Gentile life, departed more and more from their religious and social traditions, and adopted

Gentile customs, manners, and habits of thought. It seemed as though they were completely losing their identity in the superior numbers and more highly organized culture of the nations among whom they lived, so that in a few generations there would be no trace of them left. A complete disappearance of Jewish nationality in Central and Western Europe seemed inevitable.

But events turned out otherwise. Nationalities of different race seem to have an instinct which prevents them from fusing. However much the Jews adapted themselves, in language, manners, and to a great extent even in the forms of religion, to the European peoples among whom they lived, the feeling of strangeness between the Jews and their hosts never disappeared. This spontaneous feeling is the ultimate cause of anti-Semitism, which is, therefore, not to be got rid of by well-meaning propaganda. Nationalities want to pursue their own path, not to blend. A satisfactory state of affairs can only be brought about by mutual toleration and respect.

The first step in that direction is that we Jews should once more become conscious of our existence as a nationality and regain the self-respect that is necessary to a healthy existence. We must learn once more to glory in our ancestors and our history and once again take upon ourselves, as a nation, cultural tasks of a sort calculated to strengthen our sense of the community. It is not enough for us to play a part as individuals in the cultural development of the human race; we must also tackle tasks which only nations as a whole can perform. Only so can the Jews regain social health.

It is from this point of view that I would have you look at the Zionist movement. Today history has assigned to us the task of taking an active part in the economic and cultural reconstruction of our native land. Enthusiasts, men of brilliant gifts, have cleared the way, and many excellent members of our race are prepared to devote themselves heart and soul to the cause. May every one of them fully realize the importance of this work and contribute, according to his powers, to its success!

WORKING PALESTINE

Mein Weltbild, *Amsterdam: Querido Verlag, 1934.*

Among Zionist organizations "Working Palestine" is the one whose work is of most direct benefit to the most valuable class of people living there, namely, those who are transforming deserts into flourishing settlements by the labor of their hands. These workers are a selection, made on the voluntary basis, from the whole Jewish nation, an élite composed of strong, confident, and unselfish people. They are not ignorant laborers who sell the labor of their hands to the highest bidder, but educated, intellectually vigorous, free men, from whose peaceful struggle with a neglected soil the whole Jewish nation are the gainers, directly and indirectly. By lightening their heavy lot as far as we can we shall be saving the most valuable sort of human life; for the first settlers' struggle on ground not yet made habitable is a difficult and dangerous business involving a heavy personal sacrifice. How true this is, only they can judge who have seen it with their own eyes. Anyone who helps to improve the equipment of these men is helping on the good work at a crucial point.

It is, moreover, this working class alone that has the power to establish healthy relations with the Arabs, which is the most important political task of Zionism. Administrations come and go; but it is human relations that finally tune the scale in the lives of nations. Therefore to support "Working Palestine" is at the same time to promote a humane and worthy policy in Palestine and to oppose an effective resistance to those undercurrents of narrow nationalism from which the whole political world, and in a less degree the small political world of Palestine affairs, is suffering.

JEWISH RECOVERY

Mein Weltbild, *Amsterdam: Querido Verlag, 1934.*

I gladly accede to your paper's request that I should address an appeal to the Jews of Hungary on behalf of Keren Hajessod.

The greatest enemies of the national consciousness and honor of the Jews are fatty degeneration—by which I mean the unconscionableness which comes from wealth and ease—and a kind of inner dependence on the surrounding Gentile world which has grown out of the loosening of the fabric of Jewish society. The best in man can only flourish when he loses himself in a community. Hence the moral danger of the Jew who has lost touch with his own people and is regarded as a foreigner by the people of his adoption. Only too often a contemptible and joyless egoism has resulted from such circumstances. The weight of outward oppression on the Jewish people is particularly heavy at the moment. But this very bitterness has done us good. A revival of Jewish national life, such as the last generation could never have dreamed of, has begun. Through the operation of a newly awakened sense of solidarity among the Jews, the scheme of colonizing Palestine, launched by a handful of devoted and judicious leaders in the face of apparently insuperable difficulties, has already prospered so far that I feel no doubt about its permanent success. The value of this achievement for the Jews everywhere is very great. Palestine will be a center of culture for all Jews, a refuge for the most grievously oppressed, a field of action for the best among us, a unifying ideal, and a means of attaining inward health for the Jews of the whole world.

CHRISTIANITY AND JUDAISM

Mein Weltbild, *Amsterdam: Querido Verlag, 1934.*

If one purges the Judaism of the Prophets and Christianity as Jesus Christ taught it of all subsequent additions, especially those of the priests, one is left with a teaching which is capable of curing all the social ills of humanity.

It is the duty of every man of good will to strive steadfastly in his own little world to make this teaching of pure humanity a living force, so far as he can. If he makes an honest attempt in this direction without being crushed and trampled underfoot by his contemporaries, he may consider himself and the community to which he belongs lucky.

JEWISH IDEALS

Mein Weltbild, *Amsterdam: Querido Verlag, 1934.*

The pursuit of knowledge for its own sake, an almost fanatical love of justice and the desire for personal independence— these are the features of the Jewish tradition which make me thank my stars that I belong to it.

Those who are raging today against the ideals of reason and individual liberty and are trying to establish a spiritless state-slavery by brute force rightly see in us their irreconcilable foes. History has given us a difficult row to hoe; but so long as we remain devoted servants of truth, justice, and liberty, we shall continue not merely to survive as the oldest of living peoples, but by creative work to bring forth fruits which contribute to the ennoblement of the human race, as heretofore.

IS THERE A JEWISH POINT OF VIEW?

Mein Weltbild, *Amsterdam: Querido Verlag, 1934.*

In the philosophical sense there is, in my opinion, no specifically Jewish point of view. Judaism seems to me to be concerned almost exclusively with the moral attitude in life and to life. I look upon it as the essence of an attitude to life which is incarnate in the Jewish people rather than the essence of the

laws laid down in the Torah and interpreted in the Talmud.
To me, the Torah and the Talmud are merely the most important evidence of the manner in which the Jewish concept of life
held sway in earlier times.

The essence of that conception seems to me to lie in an
affirmative attitude to the life of all creation. The life of the
individual only has meaning in so far as it aids in making the
life of every living thing nobler and more beautiful. Life is
sacred, that is to say, it is the supreme value, to which all other
values are subordinate. The hallowing of the supra-individual
life brings in its train a reverence for everything spiritual—a
particularly characteristic feature of the Jewish tradition.

Judaism is not a creed: the Jewish God is simply a negation
of superstition, an imaginary result of its elimination. It is
also an attempt to base the moral law on fear, a regrettable and
discreditable attempt. Yet it seems to me that the strong moral
tradition of the Jewish nation has to a large extent shaken itself
free from this fear. It is clear also that "serving God" was
equated with "serving the living." The best of the Jewish
people, especially the Prophets and Jesus, contended tirelessly
for this.

Judaism is thus no transcendental religion; it is concerned
with life as we live it and as we can, to a certain extent, grasp
it, and nothing else. It seems to me, therefore, doubtful whether
it can be called a religion in the accepted sense of the word,
particularly as no "faith" but the sanctification of life in a supra-
personal sense is demanded of the Jew.

But the Jewish tradition also contains something else, something which finds splendid expression in many of the Psalms,
namely, a sort of intoxicated joy and amazement at the beauty
and grandeur of this world, of which man can form just a faint
notion. This joy is the feeling from which true scientific research draws its spiritual sustenance, but which also seems to
find expression in the song of birds. To tack this feeling to the
idea of God seems mere childish absurdity.

Is what I have described a distinguishing mark of Judaism?
Is it to be found anywhere else under another name? In its

pure· form, it is nowhere to be found, not even in Judaism, where the pure doctrine is obscured by much worship of the letter. Yet Judaism seems to me one of its purest and most vigorous manifestations. This applies particularly to the fundamental principle of the sanctification of life.

It is characteristic that the animals were expressly included in the command to keep holy the Sabbath day, so strong was the feeling of the ideal solidarity of all living things. The insistence on the solidarity of all human beings finds still stronger expression, and it is no mere chance that the demands of Socialism were tor the most part first raised by Jews.

How strongly developed this sense of the sanctity of life is in the Jewish people is admirably illustrated by a little remark which Walter Rathenau once made to me in conversation: "When a Jew says that he's going hunting to amuse himself, he lies." The Jewish sense of the sanctity of life could not be more simply expressed.

ANTI-SEMITISM AND ACADEMIC YOUTH

Mein Weltbild, *Amsterdam: Querido Verlag, 1934.*

So long as we lived in the ghetto our Jewish nationality involved us in material difficulties and sometimes physical danger, but no social or psychological problems. With emancipation the position changed, particularly for those Jews who turned to the intellectual professions.

In school and at the university the young Jew is exposed to the influence of a society which has a definite national tinge, which he respects and admires, from which he receives his mental sustenance, and to which he feels himself to belong; while on the other hand this society treats him, as one of an alien race, with a certain contempt and hostility. Driven by the suggestive influence·of this psychological superiority rather than by utilitarian considerations, he turns his back on his people and his traditions, and chooses to consider himself as belonging entirely to the others; the while he tries in vain to conceal from himself and them the fact that the relation is not reciprocal. Hence that pathetic creature, the baptized Jewish *Geheimrat* of yesterday and today. In most cases it is not pushfulness and lack of character that have made him what he is but, as I have said, the suggestive power of an environment superior in numbers and influence. He knows, of course, that many admirable sons of the Jewish people have made important contributions to the glory of European civilization; but have they not all, with a few exceptions, done much the same as he?

In this case, as in many mental disorders, the cure lies in a clear knowledge of one's condition and its causes. We must be conscious of our alien race and draw the logical conclusions from it. It is no use trying to convince the others of our spiritual and intellectual equality by arguments addressed to the reason, when the attitude of these others does not originate in their intellects at all. Rather must we emancipate ourselves socially, and supply our social needs, in the main, ourselves. We must have our own students' societies and adopt an attitude of courteous but consistent reserve to the Gentiles. And let us live after our own fashion there and not ape dueling and drinking customs which are foreign to our nature. It is possible to be a civilized European and a good citizen and at the same time a faithful Jew who loves his race and honors his fathers. If we remember this and act accordingly, the problem of anti-Semitism, in so far as it is of a social nature, is solved for us.

OUR DEBT TO ZIONISM

From an address on the occasion of the celebration of the "Third Seder" by the National Labor Committee for Palestine, at the Commodore Hotel in New York City, April 17, 1938. Published in New Palestine, *Washington, D. C.: April 28, 1938.*

Rarely since the conquest of Jerusalem by Titus has the Jewish community experienced a period of greater oppression than prevails at the present time. In some respects, indeed, our own time is even more troubled, for man's possibilities of emigration are more limited today than they were then.

Yet we shall survive this period, too, no matter how much sorrow, no matter how heavy a loss in life it may bring. A community like ours, which is a community purely by reason of tradition, can only be strengthened by pressure from without. For today every Jew feels that to be a Jew means to bear a serious responsibility not only to his own community, but also toward humanity. To be a Jew, after all, means first of all, to acknowledge and follow in practice those fundamentals in humaneness laid down in the Bible—fundamentals without which no sound and happy community of men can exist.

We meet today because of our concern for the development of Palestine. In this hour one thing, above all, must be emphasized: Judaism owes a great debt of gratitude to Zionism. The Zionist movement has revived among Jews the sense of community. It has performed productive work surpassing all the expectations anyone could entertain. This productive work in Palestine, to which self-sacrificing Jews throughout the world have contributed, has saved a large number of our brethren from direst need. In particular, it has been possible to lead a not inconsiderable part of our youth toward a life of joyous and creative work.

Now the fateful disease of our time—exaggerated nationalism, borne up by blind hatred—has brought our work in Palestine to a most difficult stage. Fields cultivated by day must have armed protection at night against fanatical Arab outlaws. All economic life suffers from insecurity. The spirit of enterprise languishes and a certain measure of unemployment (modest when measured by American standards) has made its appearance.

The solidarity and confidence with which our brethren in Palestine face these difficulties deserve our admiration. Voluntary contributions by those still employed keep the unemployed above water. Spirits remain high, in the conviction that reason and calm will ultimately reassert themselves. Everyone knows that the riots are artificially fomented by those directly interested in embarrassing not only ourselves but especially England. Everyone knows that banditry would cease if foreign subsidies were withdrawn.

Our brethren in other countries, however, are in no way behind those in Palestine. They, too, will not lose heart but will resolutely and firmly stand behind the common work. This goes without saying.

Just one more personal word on the question of partition. I should much rather see reasonable agreement with the Arabs on the basis of living together in peace than the creation of a Jewish state. Apart from practical consideration, my awareness of the essential nature of Judaism resists the idea of a Jewish state with borders, an army, and a measure of temporal power no matter how modest. I am afraid of the inner damage Judaism will sustain—especially from the development of a narrow nationalism within our own ranks, against which we have already had to fight strongly, even without a Jewish state. We are no longer the Jews of the Maccabee period. A return to a nation in the political sense of the word would be equivalent to turning away from the spiritualization of our community which we owe to the genius of our prophets. If external necessity should after all compel us to assume this burden, let us bear it with tact and patience.

One more word on the present psychological attitude of the world at large, upon which our Jewish destiny also depends. Anti-Semitism has always been the cheapest means employed by selfish minorities for deceiving the people. A tyranny based on such deception and maintained by terror must inevitably perish from the poison it generates within itself. For the pressure of accumulated injustice strengthens those moral forces in man which lead to a liberation and purification of public life. May our community through its suffering and its work contribute toward the release of those liberating forces.

WHY DO THEY HATE THE JEWS?

From Collier's Magazine, *New York, November 26, 1938.*

I should like to begin by telling you an ancient fable, with a few minor changes—a fable that will serve to throw into bold relief the mainsprings of political anti-Semitism:

The shepherd boy said to the horse: "You are the noblest beast that treads the earth. You deserve to live in untroubled bliss; and indeed your happiness would be complete were it not for the treacherous stag. But he practiced from youth to excel you in fleetness of foot. His faster pace allows him to reach the water holes before you do. He and his tribe drink up the water far and wide, while you and your foal are left to thirst. Stay with me! My wisdom and guidance shall deliver you and your kind from a dismal and ignominious state."

Blinded by envy and hatred of the stag, the horse agreed. He yielded to the shepherd lad's bridle. He lost his freedom and became the shepherd's slave.

The horse in this fable represents a people, and the shepherd lad a class or clique aspiring to absolute rule over the people; the stag, on the other hand, represents the Jews.

I can hear you say: "A most unlikely tale! No creature would be as foolish as the horse in your fable." But let us give it a little more thought. The horse had been suffering the pangs of thirst, and his vanity was often pricked when he saw the nimble stag outrunning him. You, who have known no such pain and vexation, may find it difficult to understand that hatred and blindness should have driven the horse to act with such ill-advised, gullible haste. The horse, however, fell an easy victim to temptation because his earlier tribulations had prepared him for such a blunder. For there is much truth in the saying that it is easy to give just and wise counsel—to others!—but hard to act justly and wisely for oneself. I say to you with full conviction: We all have often played the tragic rôle of the horse and we are in constant danger of yielding to temptation again.

The situation illustrated in this fable happens again and again in the life of individuals and nations. In brief, we may call it the process by which dislike and hatred of a given person or group are diverted to another person or group incapable of effective defense. But why did the rôle of the stag in the fable so often fall to the Jews? Why did the Jews so often happen to draw the hatred of the masses? Primarily because there are Jews among almost all nations and because they are everywhere too thinly scattered to defend themselves against violent attack.

A few examples from the recent past will prove the point: Toward the end of the nineteenth century the Russian people were chafing under the tyranny of their government. Stupid blunders in foreign policy further strained their temper until it reached the breaking point. In this extremity the rulers of Russia sought to divert unrest by inciting the masses to hatred and violence toward the Jews. These tactics were repeated after the Russian government had drowned the dangerous revolution of 1905 in blood—and this maneuver may well have helped to keep the hated regime in power until near the end of the World War.

When the Germans had lost the World War hatched by their ruling class, immediate attempts were made to blame the Jews, first for instigating the war and then for losing it. In the course of time, success attended these efforts. The hatred engendered against the Jews not only protected the privileged classes, but enabled a small, unscrupulous, and insolent group to place the German people in a state of complete bondage.

The crimes with which the Jews have been charged in the course of history—crimes which were to justify the atrocities perpetrated against them—have changed in rapid succession. They were supposed to have poisoned wells. They were said to have murdered children for ritual purposes. They were falsely charged with a systematic attempt at the economic domination and exploitation of all mankind. Pseudo-scientific books were written to brand them an inferior, dangerous race. They were reputed to foment wars and revolutions for their own selfish purposes. They were presented at once as dangerous innovators and as enemies of true progress. They were charged with falsifying the culture of nations by penetrating the national life under the guise of becoming assimilated. In the same breath they were accused of being so stubbornly inflexible that it was impossible for them to fit into any society.

Almost beyond imagination were the charges brought against them, charges known to their instigators to be untrue all the while, but which time and again influenced the masses. In times of unrest and turmoil the masses are inclined to hatred and cruelty, whereas in times of peace these traits of human nature emerge but stealthily.

Up to this point I have spoken only of violence and oppression against the Jews—not of anti-Semitism itself as a psychological and social phenomenon existing even in times and circumstances when no special action against the Jews is under way. In this sense, one may speak of latent anti-Semitism. What is its basis? I believe that in a certain sense one may actually regard it as a normal manifestation in the life of a people.

The members of any group existing in a nation are more closely bound to one another than they are to the remaining population. Hence a nation will never be free of friction while such groups continue to be distinguishable. In my belief, uniformity in a population would not be desirable, even if it were attainable. Common convictions and aims, similar interests, will in every society produce groups that, in a certain sense, act as units. There will always be friction between such groups —the same sort of aversion and rivalry that exists between individuals.

The need for such groupings is perhaps most easily seen in the field of politics, in the formation of political parties. Without parties the political interests of the citizens of any state are bound to languish. There would be no forum for the free exchange of opinions. The individual would be isolated and unable to assert his convictions. Political convictions, moreover, ripen and grow only through mutual stimulation and criticism offered by individuals of similar disposition and purpose; and politics is no different from any other field of our cultural existence. Thus it is recognized, for example, that in times of intense religious fervor different sects are likely to spring up whose rivalry stimulates religious life in general. It is well known, on the other hand, that centralization—that is, elimination of independent groups—leads to one-sidedness and barrenness in science and art because such centralization checks and even suppresses any rivalry of opinions and research trends.

Just What Is a Jew?

The formation of groups has an invigorating effect in all spheres of human striving, perhaps mostly due to the struggle between the convictions and aims represented by the different groups. The Jews, too, form such a group with a definite character of its own, and anti-Semitism is nothing but the antagonistic attitude produced in the non-Jews by the Jewish group. This is a normal social reaction. But for the political abuse resulting from it, it might never have been designated by a special name.

What are the characteristics of the Jewish group? What, in the first place, is a Jew? There are no quick answers to this question. The most obvious answer would be the following: A Jew is a person professing the Jewish faith. The superficial character of this answer is easily recognized by means of a simple parallel. Let us ask the question: What is a snail? An answer similar in kind to the one given above might be: A snail is an animal inhabiting a snail shell. This answer is not altogether incorrect; nor, to be sure, is it exhaustive; for the snail shell happens to be but one of the material products of the snail. Similarly, the Jewish faith is but one of the characteristic products of the Jewish community. It is, furthermore, known that a snail can shed its shell without thereby ceasing to be a snail. The Jew who abandons his faith (in the formal sense of the word) is in a similar position. He remains a Jew.

Difficulties of this kind appear whenever one seeks to explain the essential character of a group.

The bond that has united the Jews for thousands of years and that unites them today is, above all, the democratic ideal of social justice, coupled with the ideal of mutual aid and tolerance among all men. Even the most ancient religious scriptures of the Jews are steeped in these social ideals, which have powerfully affected Christianity and Mohammedanism and have had a benign influence upon the social structure of a great part of mankind. The introduction of a weekly day of rest should be remembered here—a profound blessing to all mankind. Personalities such as Moses, Spinoza, and Karl Marx, dissimilar as they may be, all lived and sacrificed themselves for the ideal of social justice; and it was the tradition of their forefathers that led them on this thorny path. The unique accomplishments of the Jews in the field of philanthropy spring from the same source.

The second characteristic trait of Jewish tradition is the high regard in which it holds every form of intellectual aspiration and spiritual effort. I am convinced that this great respect for intellectual striving is solely responsible for the contributions that the Jews have made toward the progress of knowledge, in the broadest sense of the term. In view of their relatively small number and the considerable external obstacles constantly placed in their way on all sides, the extent of those contributions deserves the admiration of all sincere men. I am convinced that this is not due to any special wealth of endowment, but to the fact that the esteem in which intellectual accomplishment is held among the Jews creates an atmosphere particularly favorable to the development of any talents that may exist. At the same time a strong critical spirit prevents blind obeisance to any mortal authority.

I have confined myself here to these two traditional traits, which seem to me the most basic. These standards and ideals find expression in small things as in large. They are transmitted from parents to children; they color conversation and judgment among friends; they fill the religious scriptures; and they give to the community life of the group its characteristic stamp. It is in these distinctive ideals that I see the essence of Jewish na-

ture. That these ideals are but imperfectly realized in the group
—in its actual everyday life—is only natural. However, if one
seeks to give brief expression to the essential character of a
group, the approach must always be by the way of the ideal.

WHERE OPPRESSION IS A STIMULUS

In the foregoing I have conceived of Judaism as a community
of tradition. Both friend and foe, on the other hand, have often
asserted that the Jews represent a race; that their characteristic
behavior is the result of innate qualities transmitted by *heredity*
from one generation to the next. This opinion gains weight
from the fact that the Jews for thousands of years have predominantly married within their own group. Such a custom may indeed *preserve* a homogeneous race—if it existed originally; it
cannot *produce* uniformity of the race—if there was originally
a racial intermixture. The Jews, however, are beyond doubt
a mixed race, just as are all other groups of our civilization.
Sincere anthropologists are agreed on this point; assertions to
the contrary all belong to the field of political propaganda and
must be rated accordingly.

Perhaps even more than on its own tradition, the Jewish
group has thrived on oppression and on the antagonism it has
forever met in the world. Here undoubtedly lies one of the
main reasons for its continued existence through so many thousands of years.

The Jewish group, which we have briefly characterized in
the foregoing, embraces about sixteen million people—less than
one per cent of mankind, or about half as many as the population of present-day Poland. Their significance as a political factor is negligible. They are scattered over almost the entire earth
and are in no way organized as a whole—which means that they
are incapable of concerted action of any kind.

Were anyone to form a picture of the Jews solely from the
utterances of their enemies, he would have to reach the conclusion that they represent a world power. At first sight that
seems downright absurd; and yet, in my view, there is a certain

meaning behind it. The Jews as a group may be powerless, but the sum of the achievements of their individual members is everywhere considerable and telling, even though these achievements were made in the face of obstacles. The forces dormant in the individual are mobilized, and the individual himself is stimulated to self-sacrificing effort, by the spirit that is alive in the group.

Hence the hatred of the Jews by those who have reason to shun popular enlightenment. More than anything else in the world, they fear the influence of men of intellectual independence. I see in this the essential cause for the savage hatred of Jews raging in present-day Germany. To the Nazi group the Jews are not merely a means for turning the resentment of the people away from themselves, the oppressors; they see the Jews as a nonassimilable element that cannot be driven into uncritical acceptance of dogma, and that, therefore—as long as it exists at all—threatens their authority because of its insistence on popular enlightenment of the masses.

Proof that this conception goes to the heart of the matter is convincingly furnished by the solemn ceremony of the burning of the books staged by the Nazi regime shortly after its seizure of power. This act, senseless from a political point of view, can only be understood as a spontaneous emotional outburst. For that reason it seems to me more revealing than many acts of greater purpose and practical importance.

In the field of politics and social science there has grown up a justified distrust of generalizations pushed too far. When thought is too greatly dominated by such generalizations, misinterpretations of specific sequences of cause and effect readily occur, doing injustice to the actual multiplicity of events. Abandonment of generalization, on the other hand, means to relinquish understanding altogether. For that reason I believe one may and must risk generalization, as long as one remains aware of its uncertainty. It is in this spirit that I wish to present in all modesty my conception of anti-Semitism, considered from a general point of view.

In political life I see two opposed tendencies at work, locked in constant struggle with each other. The first, optimistic trend proceeds from the belief that the free unfolding of the productive forces of individuals and groups essentially leads to a satisfactory state of society. It recognizes the need for a central power, placed above groups and individuals, but concedes to such power only organizational and regulatory functions. The second, pessimistic trend assumes that free interplay of individuals and groups leads to the destruction of society; it thus seeks to base society exclusively upon authority, blind obedience, and coercion. Actually this trend is pessimistic only to a limited extent: for it is optimistic in regard to those who are, and desire to be, the bearers of power and authority. The adherents of this second trend are the enemies of the free groups and of education for independent thought. They are, moreover, the carriers of political anti-Semitism.

Here in America all pay lip service to the first, optimistic, tendency. Nevertheless, the second group is strongly represented. It appears on the scene everywhere, though for the most part it hides its true nature. Its aim is political and spiritual dominion over the people by a minority, by the circuitous route of control over the means of production. Its proponents have already tried to utilize the weapon of anti-Semitism as well as of hostility to various other groups. They will repeat the attempt in times to come. So far all such tendencies have failed because of the people's sound political instinct.

And so it will remain in the future, if we cling to the rule: Beware of flatterers, especially when they come preaching hatred.

THE DISPERSAL OF EUROPEAN JEWRY

From an address by radio for the United Jewish Appeal, broadcast March 22, 1939. Published in Out of My Later Years, *New York: Philosophical Library, 1950.*

The history of the persecutions which the Jewish people have had to suffer is almost inconceivably long. Yet the war that is being waged against us in Central Europe today falls into a special category of its own. In the past we were persecuted *despite* the fact that we were the people of the Bible; today, however, it is just *because* we are the people of the Book that we are persecuted. The aim is to exterminate not only ourselves but to destroy, together with us, that spirit expressed in the Bible and in Christianity which made possible the rise of civilization in Central and Northern Europe. If this aim is achieved, Europe will become a barren waste. For human community life cannot long endure on a basis of crude force, brutality, terror, and hate.

Only understanding for our neighbors, justice in our dealings, and willingness to help our fellow men can give human society permanence and assure security for the individual. Neither intelligence nor inventions nor institutions can serve as substitutes for these most vital parts of education.

Many Jewish communities have been uprooted in the wake of the present upheaval in Europe. Hundreds of thousands of men, women, and children have been driven from their homes and made to wander in despair over the highways of the world. The tragedy of the Jewish people today is a tragedy which reflects a challenge to the fundamental structure of modern civilization.

One of the most tragic aspects of the oppression of Jews and other groups has been the creation of a refugee class. Many distinguished men in science, art, and literature have been driven from the lands which they enriched with their talents. In a period of economic decline these exiles have within them the possibilities for reviving economic and cultural effort; many of these refugees are highly skilled experts in industry and science. They have a valuable contribution to make to the progress of the world. They are in a position to repay hospitality with new economic development and the opening up of new opportunities of employment for native populations. I am told that in England the admission of refugees was directly responsible for giving jobs to 15,000 unemployed.

As one of the former citizens of Germany who have been fortunate enough to leave that country, I know I can speak for my fellow refugees, both here and in other countries, when I give thanks to the democracies of the world for the splendid manner in which they have received us. We, all of us, owe a debt of gratitude to our new countries, and each and every one of us is doing the utmost to show our gratitude by the quality of our contributions to the economic, social, and cultural work of the countries in which we reside.

It is, however, a source of gravest concern that the ranks of the refugees are being constantly increased. The developments of the past week have added several hundred thousand potential refugees from Czechoslovakia. Again we are confronted with a major tragedy for a Jewish community which had a noble tradition of democracy and communal service.

The power of resistance which has enabled the Jewish people to survive for thousands of years is a direct outgrowth of Jewish adherence to the Biblical doctrines on the relationships among men. In these years of affliction our readiness to help one another is being put to an especially severe test. Each of us must personally face this test, that we may stand it as well as our fathers did before us. We have no other means of self-defense than our solidarity and our knowledge that the cause for which we are suffering is a momentous and sacred cause.

THE JEWS OF ISRAEL

From a radio broadcast for the United Jewish Appeal, November 27, 1949. Published in Out of My Later Years, *New York: Philosophical Library, 1950.*

There is no problem of such overwhelming importance to us Jews as consolidating that which has been accomplished in Israel with amazing energy and an unequaled willingness for sacrifice. May the joy and admiration that fill us when we think of all that this small group of energetic and thoughtful people has achieved give us the strength to accept the great responsibility which the present situation has placed upon us.

When appraising the achievement, however, let us not lose sight of the cause to be served by this achievement: rescue of our endangered brethren, dispersed in many lands, by uniting them in Israel; creation of a community which conforms as closely as possible to the ethical ideals of our people as they have been formed in the course of a long history.

One of these ideals is peace, based on understanding and self-restraint, and not on violence. If we are imbued with this ideal, our joy becomes somewhat mingled with sadness, because our relations with the Arabs are far from this ideal at the present time. It may well be that we would have reached this ideal, had we been permitted to work out, undisturbed by others, our relations with our neighbors, for we *want* peace and we realize that our future development depends on peace.

It was much less our own fault or that of our neighbors than of the Mandatory Power that we did not achieve an undivided Palestine in which Jews and Arabs would live as equals, free, in peace. If one nation dominates other nations, as was the case in the British Mandate over Palestine, she can hardly avoid following the notorious device of *Divide et Impera*. In plain language this means: create discord among the governed people so they will not unite in order to shake off the yoke imposed upon them. Well, the yoke has been removed, but the seed of dissension has borne fruit and may still do harm for some time to come—let us hope not for too long.

The Jews of Palestine did not fight for political independence for its own sake, but they fought to achieve free immigration for the Jews of many countries where their very existence was in danger; free immigration also for all those who were longing for a life among their own. It is no exaggeration to say that they fought to make possible a sacrifice perhaps unique in history.

I do not speak of the loss in lives and property fighting an opponent who was numerically far superior, nor do I mean the exhausting toil which is the pioneer's lot in a neglected arid country. I am thinking of the additional sacrifice that a population living under such conditions has to make in order to receive, in the course of eighteen months, an influx of immigrants who comprise more than one-third of the total Jewish population of the country. In order to realize what this means you have only to visualize a comparable feat of the American Jews. Let us assume there were no laws limiting the immigration into the United States; imagine that the Jews of this country volunteered to receive more than one million Jews from other countries in the course of one year and a half, to take care of them, and to integrate them into the economy of this country. This would be a tremendous achievement, but still very far from the achievement of our brethren in Israel. For the United States is a big, fertile country, sparsely populated, with a high living standard and a highly developed productive capacity, not to compare with small Jewish Palestine whose inhabitants, even without the additional burden of mass immigration, lead a hard and frugal life, still threatened by enemy attacks. Think of the privations and personal sacrifices which this voluntary act of brotherly love means for the Jews of Israel.

The economic means of the Jewish Community in Israel do not suffice to bring this tremendous enterprise to a successful end. For a hundred thousand out of more than three hundred thousand persons who immigrated to Israel since May, 1948, no homes or work could be made available. They had to be concentrated in improvised camps under conditions which are a disgrace to all of us.

It must not happen that this magnificent work breaks down because the Jews of this country do not help sufficiently or quickly enough. Here, to my mind, is a precious gift with which all Jews have been presented: the opportunity to take an active part in this wonderful task.

PART IV

ON GERMANY

MANIFESTO—MARCH, 1933

Mein Weltbild, *Amsterdam: Querido Verlag, 1934.*

As long as I have any choice, I will only stay in a country where political liberty, tolerance, and equality of all citizens before the law prevail. Political liberty implies the freedom to express one's political opinions orally and in writing; tolerance implies respect for any and every individual opinion.

These conditions do not obtain in Germany at the present time. Those who have done most for the cause of international understanding, among them some of the leading artists, are being persecuted there.

Any social organism can become psychically distempered just as any individual can, especially in times of difficulty. Nations usually survive these distempers. I hope that healthy conditions will soon supervene in Germany and that in future her great men like Kant and Goethe will not merely be commemorated from time to time but that the principles which they taught will also prevail in public life and in the general consciousness.

CORRESPONDENCE WITH THE PRUSSIAN ACADEMY OF SCIENCES

Mein Weltbild, *Amsterdam: Querido Verlag, 1934.*

THE ACADEMY'S DECLARATION OF APRIL 1, 1933 AGAINST EINSTEIN

The Prussian Academy of Sciences heard with indignation from the newspapers of Albert Einstein's participation in the atrocity-mongering in France and America. It immediately demanded an explanation. In the meantime Einstein has announced his withdrawal from the Academy, giving as his reason that he cannot continue to serve the Prussian state under its present government. Being a Swiss citizen, he also, it seems, intends to resign the Prussian citizenship which he acquired in 1913 incidental to his becoming a full member of the Academy.

The Prussian Academy of Sciences is particularly distressed by Einstein's activities as an agitator in foreign countries, as it and its members have always felt themselves bound by the closest ties to the Prussian state and, while abstaining strictly from all political partisanship, have always stressed and remained faithful to the national idea. It has therefore no reason to regret Einstein's withdrawal.

For the Prussian Academy of Sciences
(signed)   Prof. Dr. Ernst Heymann,
Perpetual Secretary

EINSTEIN'S DECLARATION TO THE ACADEMY

Le Coq, near Ostende. April 5, 1933

I have received information from a thoroughly reliable source that the Academy of Sciences has spoken in an official statement of "Albert Einstein's participation in atrocity-mongering in America and France."

I hereby declare that I have never taken any part in atrocity-mongering, and I must add that I have seen nothing of any such mongering anywhere. In general, people have contented themselves with reproducing and commenting on the official statements and orders of responsible members of the German government, together with the program for the annihilation of the German Jews by economic methods.

The statements I have issued to the Press were concerned with my intention to resign my position in the Academy and renounce my Prussion citizenship; I gave as my reason for these steps that I did not wish to live in a country where the individual does not enjoy equality before the law, and freedom of speech and teaching.

Further, I described the present state of affairs in Germany as a state of psychic distemper in the masses and made some remarks about its causes.

In a document which I allowed the International League for Combating Anti-Semitism to make use of for the purpose of enlisting support and which was not intended for the Press at all, I also called upon all sensible people, who are still faithful to the ideals of civilization in peril, to do their utmost to prevent this mass-psychosis, which manifests itself in such terrible symptoms in Germany today, from spreading any further.

It would have been an easy matter for the Academy to get hold of a correct version of my words before issuing the sort of statement about me that it has. The German Press has reproduced a deliberately distorted version of my words, as indeed was only to be expected with the Press muzzled as it is today.

I am ready to stand by every word I have published. In return, I expect the Academy to communicate this statement of mine to its members and also to the German public before which I have been slandered, especially as it has itself had a hand in slandering me before that public.

Two Communications of the Prussian Academy

Berlin, April 7, 1933

Dear Sir:

As the present Principal Secretary of the Prussian Academy I beg to acknowledge the receipt of your communication dated March 28 announcing your resignation of your membership of the Academy.

The Academy has taken note of your resignation in its plenary session of March 30, 1933.

While the Academy profoundly regrets the turn events have taken, this regret concerns the fact that a man of the highest scientific authority, whom many years of work among Germans and many years of membership of our society must have made familiar with the German character and German habits of thought, should have chosen this moment to associate himself with a body of people abroad who—partly no doubt through ignorance of actual conditions and events—have done much damage to our German people by disseminating erroneous views and unfounded rumors. We had confidently expected that one who had belonged to our Academy for so long would have ranged himself, irrespective of his own political sympathies, on the side of the defenders of our nation against the flood of lies which has been let loose upon it. In these days of mudslinging, some of it vile, some of it ridiculous, a good word for the German people from you in particular might have produced a great effect abroad. Instead of which your testimony has served as a handle to the enemies not merely of the present Government but of the German people. This has come as a bitter and grievous disappointment to us, which would no doubt have led inevitably to a parting of the ways even if we had not received your resignation.

Yours faithfully,
(signed) von Ficker

April 11, 1933

The Academy would like to point out that its statement of April 1, 1933 was based not merely on German but principally on foreign, particularly French and Belgian, newspaper reports which Herr Einstein has not contradicted; in addition, it had before it his much canvassed statement to the League for Combating Anti-Semitism, in which he deplores Germany's relapse into the barbarism of long-passed ages. Moreover, the Academy affirms that Herr Einstein, who according to his own statement has taken no part in atrocity-mongering, has at least done nothing to counteract unjust suspicions and slanders, which, in the opinion of the Academy, it was his duty as one of its senior members to do. Instead of that Herr Einstein has made statements, and in foreign countries at that, which, coming from a man of world-wide reputation, were bound to be exploited and abused by the enemies not merely of the present German Government but of the whole German people.

For the Prussian Academy of Sciences

(*signed*) H. von Ficker }
E. Heymann } *Perpetual Secretaries*

ALBERT EINSTEIN'S ANSWER

Le Coq-sur-Mer, Belgium. April 12, 1933

I have received your communication of the 7th instant and deeply deplore the mental attitude displayed in it.

As regards the facts, I can only reply as follows: What you say about my behavior is, at bottom, merely another form of the statement you have already published, in which you accuse me of having taken part in atrocity-mongering against the German people. I have already, in my last letter, characterized this accusation as slanderous.

You have also remarked that a "good word" on my part for "the German people" would have produced a great effect abroad. To this I must reply that such a testimony as you suggest would have been equivalent to a repudiation of all those notions of justice and liberty for which I have stood all my life. Such testimony would not be, as you put it, a good word for the German people; on the contrary, it would only have helped the cause of those who are seeking to undermine the ideas and principles which have won for the German people a place of honor in the civilized world. By giving such testimony in the present circumstances I should have been contributing, even if only indirectly, to moral corruption and the destruction of all existing cultural values.

It was for this reason that I felt compelled to resign from the Academy, and your letter only shows me how right I was to do so.

CORRESPONDENCE WITH THE BAVARIAN
ACADEMY OF SCIENCES

Mein Weltbild, *Amsterdam: Querido Verlag, 1934.*

From the Academy
Munich, April 8, 1933

TO PROFESSOR ALBERT EINSTEIN

SIR:

In your letter to the Prussian Academy of Sciences you have given the present state of affairs in Germany as the reason for your resignation. The Bavarian Academy of Sciences, which some years ago elected you a corresponding member, is also a German Academy, closely allied to the Prussian and other German Academies; hence your withdrawal from the Prussian Academy of Sciences is bound to affect your relations with our Academy.

We must therefore ask you how you envisage your relations with our Academy after what has passed between yourself and the Prussian Academy.

The President of the Bavarian
Academy of Sciences

ALBERT EINSTEIN'S ANSWER

Le Coq-sur-Mer, April 21, 1933

I have given it as the reason for my resignation from the Prussian Academy that in the present circumstances I have no wish either to be a German citizen or to remain in any position of dependence on the Prussian Ministry of Education.

These reasons would not, in themselves, involve the severing of my relations with the Bavarian Academy. If I nevertheless desire my name to be removed from the list of members, it is for a different reason.

The primary duty of an Academy is to further and protect the scientific life of a country. And yet the learned societies of Germany have, to the best of my knowledge, stood by and said nothing while a not inconsiderable proportion of German scholars and students and also of academically trained professionals have been deprived of all chance of getting employment or earning a living in Germany. I do not wish to belong to any society which behaves in such a manner, even if it does so under external pressure.

A REPLY TO THE INVITATION TO PARTICIPATE IN A MEETING AGAINST ANTI-SEMITISM

The following lines are Einstein's answer to an invitation to take part in a French manifestation against anti-Semitism in Germany. Published in Mein Weltbild, *Amsterdam: Querido Verlag, 1934.*

I have considered carefully and from every angle this most important proposal, which concerns a question that I have more closely at heart than any other. As a result I have come to the conclusion that I must not take a personal part in this extremely important manifestation, for two reasons:

In the first place I am still a German citizen, and in the second I am a Jew. As regards the first point I must add that I have been active in German institutions and have always been treated with full confidence in Germany. However deeply I may regret that such horrible things are happening there, however strongly I am bound to condemn the terrible aberrations occurring with the approval of the government, it is nevertheless impossible for me to take part personally in an enterprise set under way by responsible members of a foreign government. In order that you may appreciate this fully, suppose that a French citizen in a more or less analogous situation had got up a protest against the French government's action in conjunction with prominent German statesmen. Even if you fully admitted that the protest was amply warranted by the facts, you would still, I expect, regard the behavior of your fellow citizen as an act of disloyalty. If Zola had felt it necessary to leave France at the time of the Dreyfus case, he would still certainly not have associated himself with a protest by German official personages, however much he might have approved of their action. He would have confined himself to—blushing for his countrymen.

In the second place a protest against injustice and violence is incomparably more valuable if it comes entirely from individuals who have been prompted purely by sentiments of humanity and a love of justice. This cannot be said of a man like me, a Jew who regards other Jews as his brothers. To him, an injustice done to the Jews is the same as an injustice done to himself. He must not be the judge in his own case, but wait for the judgment of impartial outsiders.

These are my reasons. But I should like to add that I have always honored and admired that highly developed sense of justice which is one of the noblest features of the French tradition.

TO THE HEROES OF THE BATTLE OF THE WARSAW GHETTO

From Bulletin of the Society of Polish Jews, *New York, 1944.*

They fought and died as members of the Jewish nation, in the struggle against organized bands of German murderers. To us these sacrifices are a strengthening of the bond between us, the Jews of all the countries. We strive to be one in suffering and in the effort to achieve a better human society, that society which our prophets have so clearly and forcibly set before us as a goal.

The Germans as an entire people are responsible for these mass murders and must be punished as a people if there is justice in the world and if the consciousness of collective responsibility in the nations is not to perish from the earth entirely. Behind the Nazi party stands the German people, who elected Hitler after he had in his book and in his speeches made his shameful intentions clear beyond the possibility of misunderstanding. The Germans are the only people who have not made any serious attempt of counteraction leading to the protection of the innocently persecuted. When they are entirely defeated and begin to lament over their fate, we must not let ourselves be deceived again, but keep in mind that they deliberately used the humanity of others to make preparation for their last and most grievous crime against humanity.

Part V

CONTRIBUTIONS TO SCIENCE

INTRODUCTION

By Valentine Bargmann, Professor of Mathematical Physics, Princeton University.

I.

The following is a brief synopsis of the development of Einstein's principal physical theories. In each case we give the date of the publication of the fundamental ideas and the date of the publication of the definitive form of the theory, leaving out the numerous—no less important—papers containing applications and refinements of the theories.

I. Theory of Relativity.
   a) Special theory.
   The first paper on the special theory of relativity (written in 1905, when Einstein was an employee of the Swiss Patent Office at Berne) presents the theory already in final form. In a second paper published a short time later Einstein drew the most important conclusion from the theory, namely, the equivalence of mass and energy, expressed in the celebrated equation, $E = m\ c^2$.
   b) General theory.
   The history of the general theory of relativity is considerably longer. In a survey of the special theory of relativity, which appeared as early as 1907, Einstein pointed out the necessity of a generalization and presented the fundamental idea that the generalization must be based on the equivalence of inertial and gravitational mass. A paper written in 1911 discusses some of the conclusions from the general theory concerning the influence of gravitation on light: (1) the influence of a gravitational field on the frequency of spectral lines (gravitational red

shift); (2) the bending of light rays by the gravitational field of the sun. (Some details were later to be modified.)

After much further work—mainly on the mathematical foundation of the theory—the definitive form of general relativity was reached and published in 1916. (By that time Einstein had already derived the third "astronomical effect" of general relativity, namely, the motion of the Mercury perihelion.)

c) Further work on the general theory.

The problems of general relativity have occupied Einstein to this day. We mention three which appear to have particular importance: (1) cosmology, (2) the problem of motion, (3) unified field theory.

(1) All of modern cosmology goes back to Einstein's paper of 1917, in which he first applied general relativity to the questions of cosmology and thereby put cosmological speculation on a firm basis. (While Einstein considered, at that time, a static universe, the later development has mostly favored the "expanding universe," in view of strong astronomical evidence. Cosmology is still actively pursued by many scientists who attempt to find a coherent theory consistent with the increasing amount of astronomical data.)

(2) General relativity was originally based on two independent hypotheses: the field equations for the gravitational field, and the law of motion for material particles. In 1927 Einstein already attacked the problem of deducing the law of motion from the field equations, and repeatedly returned to it. The definitive solution was obtained in 1949 (in collaboration with L. Infeld). Thus it was shown that the field equations alone suffice as a basis for the theory.

From the beginning, the theory of general relativity was mainly a theory of the gravitational field, in so far as the field equations for the gravitational field followed in an essentially unambiguous way from the basic ideas of general relativity. Other fields could be incorporated into the framework of general relativity in an equally unambiguous way, once their structure was known. But the connection was somewhat "loose," because general relativity could not predict either the existence or the structure of any other field (for example, that of the electromagnetic field). Therefore, several scientists (e.g., Weyl, Kaluza, Eddington) tried early to extend or to generalize the theory so as to achieve a unified theory of all fields—or at least the gravitational and the electromagnetic fields. For various reasons the early attempts were not satisfactory. Einstein himself has steadily worked on this problem since 1923, repeatedly modifying the form of the theory. The latest version was initiated in 1945 and received its definitive form in 1953 (published as Appendix II to the fourth edition of *The Meaning of Relativity*).

II. Quantum Theory.

Soon after the inauguration of the quantum theory by Max Planck in 1900, Einstein became the foremost pioneer in the new field. His first contribution appeared in the same year (1905)—and even in the same volume of the *Annalen der Physik*—as his first paper on relativity. It introduced the concept of light quanta or photons and provided the basis for much of the further work in quantum theory, in particular for Bohr's theory of the atom. In 1917 there appeared one of Einstein's most significant later papers on this subject, in which, in addition to a penetrating analysis of the properties of photons, he gave a new derivation of Planck's law of radiation based on the concept of transition probabilities. This concept has remained basic ever since.

Among Einstein's other contributions we mention the first application of the quantum theory to the theory of specific heats (1907), and the particularly important papers on the quantum theory of gases (1924–25). These introduced in full generality the new type of statistics which is now known as Bose-Einstein statistics, and also contained far-reaching ideas on electron waves, which Schroedinger credited with guiding him in his work on wave mechanics.

III. Kinetic Theory of Matter.

In the years 1902–04 Einstein wrote a series of papers in

which he independently established the theory of statistical
mechanics in a manner analogous to that of the great American
physicist, J. W. Gibbs. (Statistical mechanics or the kinetic
theory of matter derives the thermal properties of matter in
bulk from the assumption that matter consists of atoms [ulti-
mate particles] which move according to the laws of mechanics.)
The most significant sequel was a third important paper which
Einstein wrote in 1905, that on Brownian motion. In it Einstein predicted, on the basis of the kinetic theory, the motion
of minute particles suspended in a liquid. (Such a motion had
been observed about one hundred years earlier by the English
botanist, Robert Brown.) Conversely, the experimental investigation of such motions (in particular the work of the
French physicist Perrin, which was inspired by Einstein's
theory) led to a verification of the basic hypotheses of the
kinetic theory of matter.

PRINCIPLES OF THEORETICAL PHYSICS

*Inaugural address before the Prussian Academy of Sciences,
1914. Einstein became a member of the Prussian Academy
in 1913. In 1933, after the advent of the Hitler regime, he
resigned from the Academy. (See correspondence, pp. 205 ff.
of this volume.) Published in* Proceedings of the Prussian
Academy of Sciences, *1914.*

GENTLEMEN:
First of all, I have to thank you most heartily for conferring
the greatest benefit on me that anybody can confer on a man
like myself. By electing me to your Academy you have freed
me from the distractions and cares of a professional life and so
made it possible for me to devote myself entirely to scientific
studies. I beg that you will continue to believe in my gratitude
and my industry even when my efforts seem to you to yield but
a poor result.
Perhaps I may be allowed *à propos* of this to make a few

general remarks on the relation of my sphere of activity, which is theoretical physics, toward experimental physics. A mathematician friend of mine said to me the other day half in jest: "The mathematician can do a lot of things, but never what you happen to want him to do just at the moment." Much the same often applies to the theoretical physicist when the experimental physicist calls him in. What is the reason for this peculiar lack of adaptability?

The theorist's method involves his using as his foundation general postulates or "principles" from which he can deduce conclusions. His work thus falls into two parts. He must first discover his principles and then draw the conclusions which follow from them. For the second of these tasks he receives an admirable equipment at school. If, therefore, the first of his problems has already been solved for some field or for a complex of related phenomena, he is certain of success, provided his industry and intelligence are adequate. The first of these tasks, namely, that of establishing the principles which are to serve as the starting point of his deduction, is of an entirely different nature. Here there is no method capable of being learned and systematically applied so that it leads to the goal. The scientist has to worm these general principles out of nature by perceiving in comprehensive complexes of empirical facts certain general features which permit of precise formulation.

Once this formulation is successfully accomplished, inference follows on inference, often revealing unforeseen relations which extend far beyond the province of the reality from which the principles were drawn. But as long as no principles are found on which to base the deduction, the individual empirical fact is of no use to the theorist; indeed he cannot even do anything with isolated general laws abstracted from experience. He will remain helpless in the face of separate results of empirical research, until principles which he can make the basis of deductive reasoning have revealed themselves to him.

This is the kind of position in which theory finds itself at present in regard to the laws of heat radiation and molecular motion at low temperatures. About fifteen years ago nobody had yet doubted that a correct account of the electrical, optical, and thermal properties of matter was possible on the basis of Galileo-Newtonian mechanics applied to molecular motion and of Maxwell's theory of the electromagnetic field. Then Planck showed that in order to establish a law of heat radiation consonant with experience, it was necessary to employ a method of calculation whose incompatibility with the principles of classical physics became clearer and clearer. For with this method of calculation, Planck introduced into physics the quantum hypothesis, which has since received brilliant confirmation. With this quantum hypothesis he dethroned classical physics as applied to the case where sufficiently small masses move at sufficiently low speeds and sufficiently high rates of acceleration, so that today the laws of motion propounded by Galileo and Newton can only be accepted as limiting laws. In spite of assiduous efforts, however, the theorists have not yet succeeded in replacing the principles of mechanics by others which fit in with Planck's law of heat radiation or the quantum hypothesis. No matter how definitely it has been established that heat is to be explained by molecular motion, we have nevertheless to admit today that our position in regard to the fundamental laws of this motion resembles that of astronomers before Newton in regard to the motions of the planets.

I have just now referred to a group of facts for the theoretical treatment of which the principles are lacking. But it may equally well happen that clearly formulated principles lead to conclusions which fall entirely, or almost entirely, outside the sphere of reality at present accessible to our experience. In that case it may need many years of empirical research to ascertain whether the theoretical principles correspond with reality. We have an instance of this in the theory of relativity.

An analysis of the fundamental concepts of space and time has shown us that the principle of the constant velocity of light in empty space, which emerges from the optics of bodies in motion by no means forces us to accept the theory of a stationary luminiferous ether. On the contrary, it has been possible to frame a general theory which takes account of the fact that experiments carried out on the earth never reveal any trans-latory motion of the earth. This involves using the principle of relativity, which says that the laws of nature do not alter their form when one passes from the original (admissible) system of co-ordinates to a new one which is in uniform translatory motion with respect to it. This theory has received substantial confirmation from experience and has led to a simplification of the theoretical description of groups of facts already connected.

On the other hand, from the theoretical point of view this theory is not wholly satisfactory, because the principle of relativity just formulated favors *uniform* motion. If it is true that no absolute significance must be attached to *uniform* motion from the physical point of view, the question arises whether this statement must not also be extended to non-uniform motions. It has turned out that one arrives at an unambiguous extension of the relativity theory if one postulates a principle of relativity in this extended sense. One is led thereby to a general theory of gravitation which includes dynamics. For the present, however, we have not the necessary array of facts to test the legitimacy of our introduction of the postulated principle.

We have ascertained that inductive physics asks questions of deductive, and vice versa, the answers to which demand the exertion of all our energies. May we soon succeed in making permanent progress by our united efforts!

PRINCIPLES OF RESEARCH

Address delivered at a celebration of Max Planck's sixtieth birthday (1918) before the Physical Society in Berlin. Published in Mein Weltbild, *Amsterdam: Querido Verlag, 1934. Max Planck (1858–1947) was for many years professor of theoretical physics at the University of Berlin. By far the most outstanding of his contributions to physics is his quantum theory, which he advanced in 1900 and which has provided the basis for the whole development of modern atomic physics. Next to Planck it was Einstein who did the pioneering work in the young field, above all in his theory of light quanta or photons (1905) and his theory of specific heats (1907). It was he who perceived more than anyone else the fundamental and pervasive character of the quantum concept in all its ramifications.*

In the temple of science are many mansions, and various indeed are they that dwell therein and the motives that have led them thither. Many take to science out of a joyful sense of superior intellectual power; science is their own special sport to which they look for vivid experience and the satisfaction of ambition; many others are to be found in the temple who have offered the products of their brains on this altar for purely utilitarian purposes. Were an angel of the Lord to come and drive all the people belonging to these two categories out of the temple, the assemblage would be seriously depleted, but there would still be some men, of both present and past times, left inside. Our Planck is one of them, and that is why we love him.

I am quite aware that we have just now light-heartedly expelled in imagination many excellent men who are largely, perhaps chiefly, responsible for the building of the temple of science; and in many cases our angel would find it a pretty ticklish job to decide. But of one thing I feel sure: if the types we have just expelled were the only types there were, the temple would never have come to be, any more than a forest can grow which consists of nothing but creepers. For these people any sphere of human activity will do, if it comes to a point; whether they become engineers, officers, tradesmen, or scientists depends on circumstances. Now let us have another look at those who have found favor with the angel. Most of them are somewhat odd, uncommunicative, solitary fellows, really less like each other, in spite of these common characteristics, than the hosts of the rejected. What has brought them to the temple? That is a difficult question and no single answer will cover it. To begin with, I believe with Schopenhauer that one of the strongest motives that leads men to art and science is escape from everyday life with its painful crudity and hopeless dreariness, from the fetters of one's own ever shifting desires. A finely tempered nature longs to escape from personal life into the world of objective perception and thought; this desire may be compared with the townsman's irresistible longing to escape from his noisy, cramped surroundings into the silence of high mountains, where the eye ranges freely through the still, pure air and fondly traces out the restful contours apparently built for eternity.

With this negative motive there goes a positive one. Man tries to make for himself in the fashion that suits him best a simplified and intelligible picture of the world; he then tries to some extent to substitute this cosmos of his for the world of experience, and thus to overcome it. This is what the painter, the poet, the speculative philosopher, and the natural scientist do, each in his own fashion. Each makes this cosmos and its construction the pivot of his emotional life, in order to find in this way the peace and security which he cannot find in the narrow whirlpool of personal experience.

What place does the theoretical physicist's picture of the world occupy among all these possible pictures? It demands the highest possible standard of rigorous precision in the description of relations, such as only the use of mathematical language can give. In regard to his subject matter, on the other hand, the physicist has to limit himself very severely: he must content himself with describing the most simple events which can be brought within the domain of our experience; all events of a more complex order are beyond the power of the human intellect to reconstruct with the subtle accuracy and logical perfection which the theoretical physicist demands. Supreme purity, clarity, and certainty at the cost of completeness. But what can be the attraction of getting to know such a tiny section of nature thoroughly, while one leaves everything subtler and more complex shyly and timidly alone? Does the product of such a modest effort deserve to be called by the proud name of a theory of the universe?

In my belief the name is justified; for the general laws on which the structure of theoretical physics is based claim to be valid for any natural phenomenon whatsoever. With them, it ought to be possible to arrive at the description, that is to say, the theory, of every natural process, including life, by means of pure deduction, if that process of deduction were not far beyond the capacity of the human intellect. The physicist's renunciation of completeness for his cosmos is therefore not a matter of fundamental principle.

The supreme task of the physicist is to arrive at those universal elementary laws from which the cosmos can be built up by pure deduction. There is no logical path to these laws; only intuition, resting on sympathetic understanding of experience, can reach them. In this methodological uncertainty, one might suppose that there were any number of possible systems of theoretical physics all equally well justified; and this opinion is no doubt correct, theoretically. But the development of physics has shown that at any given moment, out of all conceivable constructions, a single one has always proved itself decidedly superior to all the rest. Nobody who has really gone deeply into the matter will deny that in practice the world of phenomena uniquely determines the theoretical system, in spite of the.fact that there is no logical bridge between phenomena and their theoretical principles; this is what Leibnitz described so happily as a "pre-established harmony." Physicists often accuse epistemologists of not paying sufficient attention to this fact. Here, it seems to me, lie the roots of the controversy carried on some years ago between Mach and Planck.

The longing to behold this pre-established harmony is the source of the inexhaustible patience and perseverance with which Planck has devoted himself, as we see, to the most general problems of our science, refusing to let himself be diverted to more grateful and more easily attained ends. I have often heard colleagues try to attribute this attitude of his to extraordinary will-power and discipline—wrongly, in my opinion. The state of mind which enables a man to do work of this kind is akin to that of the religious worshiper or the lover; the daily effort comes from no deliberate intention or program, but straight from the heart. There he sits, our beloved Planck, and smiles inside himself at my childish playing-about with the lantern of Diogenes. Our affection for him needs no threadbare explanation. May the love of science continue to illumine his path in the future and lead him to the solution of the most important problem in present-day physics, which he has himself posed and done so much to solve. May he succeed in uniting quantum theory with electrodynamics and mechanics in a single logical system.

WHAT IS THE THEORY OF RELATIVITY?

Written at the request of The London Times. *Published November 28, 1919.*

I gladly accede to the request of your colleague to write something for *The Times* on relativity. After the lamentable breakdown of the old active intercourse between men of learning, I welcome this opportunity of expressing my feelings of joy and gratitude toward the astronomers and physicists of England. It is thoroughly in keeping with the great and proud traditions of scientific work in your country that eminent scientists should have spent much time and trouble, and your scientific institutions have spared no expense, to test the implications of a theory which was perfected and published during the war in the land of your enemies. Even though the investigation of the influence of the gravitational field of the sun on light rays is a purely objective matter, I cannot forbear to express my personal thanks to my English colleagues for their work; for without it I could hardly have lived to see the most important implication of my theory tested.

We can distinguish various kinds of theories in physics. Most of them are constructive. They attempt to build up a picture of the more complex phenomena out of the materials of a relatively simple formal scheme from which they start out. Thus the kinetic theory of gases seeks to reduce mechanical, thermal, and diffusional processes to movements of molecules—i.e., to build them up out of the hypothesis of molecular motion. When we say that we have succeeded in understanding a group of natural processes, we invariably mean that a constructive theory has been found which covers the processes in question.

Along with this most important class of theories there exists a second, which I will call "principle-theories." These employ the analytic, not the synthetic, method. The elements which form their basis and starting-point are not hypothetically constructed but empirically discovered ones, general characteristics of natural processes, principles that give rise to mathematically formulated criteria which the separate processes or the theoretical representations of them have to satisfy. Thus the science of thermodynamics seeks by analytical means to deduce necessary conditions, which separate events have to satisfy, from the universally experienced fact that perpetual motion is impossible.

The advantages of the constructive theory are completeness, adaptability, and clearness, those of the principle theory are logical perfection and security of the foundations.

The theory of relativity belongs to the latter class. In order to grasp its nature, one needs first of all to become acquainted with the principles on which it is based. Before I go into these, however, I must observe that the theory of relativity resembles a building consisting of two separate stories, the special theory and the general theory. The special theory, on which the general theory rests, applies to all physical phenomena with the exception of gravitation; the general theory provides the law of gravitation and its relations to the other forces of nature.

It has, of course, been known since the days of the ancient Greeks that in order to describe the movement of a body, a second body is needed to which the movement of the first is referred. The movement of a vehicle is considered in reference to the earth's surface, that of a planet to the totality of the visible fixed stars. In physics the body to which events are spatially referred is called the coordinate system. The laws of the mechanics of Galileo and Newton, for instance, can only be formulated with the aid of a coordinate system.

The state of motion of the coordinate system may not, however, be arbitrarily chosen, if the laws of mechanics are to be valid (it must be free from rotation and acceleration). A coordinate system which is admitted in mechanics is called an "inertial system." The state of motion of an inertial system is according to mechanics not one that is determined uniquely by nature. On the contrary, the following definition holds good: a coordinate system that is moved uniformly and in a straight line relative to an inertial system is likewise an inertial system. By the "special principle of relativity" is meant the generalization of this definition to include any natural event whatever: thus, every universal law of nature which is valid in relation to a coordinate system C, must also be valid, as it stands, in relation to a coordinate system C', which is in uniform translatory motion relatively to C.

The second principle, on which the special theory of relativity rests, is the "principle of the constant velocity of light in vacuo." This principle asserts that light in vacuo always has a definite velocity of propagation (independent of the state of motion of the observer or of the source of the light). The confidence which physicists place in this principle springs from the successes achieved by the electrodynamics of Maxwell and Lorentz.

Both the above-mentioned principles are powerfully supported by experience, but appear not to be logically reconcilable. The special theory of relativity finally succeeded in reconciling them logically by a modification of kinematics—i.e., of the doctrine of the laws relating to space and time (from the point of view of physics). It became clear that to speak of the simultaneity of two events had no meaning except in relation to a given coordinate system, and that the shape of measuring devices and the speed at which clocks move depend on their state of motion with respect to the coordinate system.

But the old physics, including the laws of motion of Galileo and Newton, did not fit in with the suggested relativist kinematics. From the latter, general mathematical conditions issued, to which natural laws had to conform, if the above-mentioned two principles were really to apply. To these, physics had to be adapted. In particular, scientists arrived at a new law of motion for (rapidly moving) mass points, which was admirably confirmed in the case of electrically charged particles. The most important upshot of the special theory of relativity concerned the inert masses of corporeal systems. It turned out that the inertia of a system necessarily depends on its energy-content, and this led straight to the notion that inert mass is simply latent energy. The principle of the conservation of mass lost its independence and became fused with that of the conservation of energy.

The special theory of relativity, which was simply a systematic development of the electrodynamics of Maxwell and Lorentz, pointed beyond itself, however. Should the independence of physical laws of the state of motion of the coordinate system be restricted to the uniform translatory motion of coordinate systems in respect to each other? What has nature to do with our coordinate systems and their state of motion? If it is necessary for the purpose of describing nature, to make use of a coordinate system arbitrarily introduced by us, then the choice of its state of motion ought to be subject to no restriction; the laws ought to be entirely independent of this choice (general principle of relativity).

The establishment of this general principle of relativity is made easier by a fact of experience that has long been known, namely, that the weight and the inertia of a body are controlled by the same constant (equality of inertial and gravitational mass). Imagine a coordinate system which is rotating uniformly with respect to an inertial system in the Newtonian manner. The centrifugal forces which manifest themselves in relation to this system must, according to Newton's teaching, be regarded as effects of inertia. But these centrifugal forces are, exactly like the forces of gravity, proportional to the masses of the bodies. Ought it not to be possible in this case to regard the coordinate system as stationary and the centrifugal forces as gravitational forces? This seems the obvious view, but classical mechanics forbid it.

This hasty consideration suggests that a general theory of relativity must supply the laws of gravitation, and the consistent following up of the idea has justified our hopes.

But the path was thornier than one might suppose, because it demanded the abandonment of Euclidean geometry. This is to say, the laws according to which solid bodies may be arranged in space do not completely accord with the spatial laws attributed to bodies by Euclidean geometry. This is what we mean when we talk of the "curvature of space." The fundamental concepts of the "straight line," the "plane," etc., thereby lose their precise significance in physics.

In the general theory of relativity the doctrine of space and time, or kinematics, no longer figures as a fundamental independent of the rest of physics. The geometrical behavior of bodies and the motion of clocks rather depend on gravitational fields, which in their turn are produced by matter.

The new theory of gravitation diverges considerably, as regards principles, from Newton's theory. But its practical results agree so nearly with those of Newton's theory that it is difficult to find criteria for distinguishing them which are accessible to experience. Such have been discovered so far:

1. In the revolution of the ellipses of the planetary orbits round the sun (confirmed in the case of Mercury).
2. In the curving of light rays by the action of gravitational fields (confirmed by the English photographs of eclipses).

3. In a displacement of the spectral lines toward the red end of the spectrum in the case of light transmitted to us from stars of considerable magnitude (unconfirmed so far).*

The chief attraction of the theory lies in its logical completeness. If a single one of the conclusions drawn from it proves wrong, it must be given up; to modify it without destroying the whole structure seems to be impossible.

Let no one suppose, however, that the mighty work of Newton can really be superseded by this or any other theory. His great and lucid ideas will retain their unique significance for all time as the foundation of our whole modern conceptual structure in the sphere of natural philosophy.

Note: Some of the statements in your paper concerning my life and person owe their origin to the lively imagination of the writer. Here is yet another application of the principle of relativity for the delectation of the reader: today I am described in Germany as a "German savant," and in England as a "Swiss Jew." Should it ever be my fate to be represented as a *bête noire,* I should, on the contrary, become a "Swiss Jew" for the Germans and a "German savant" for the English.

GEOMETRY AND EXPERIENCE

Lecture before the Prussian Academy of Sciences, January 27, 1921. The last part appeared first in a reprint by Springer, Berlin, 1921.

One reason why mathematics enjoys special esteem, above all other sciences, is that its propositions are absolutely certain and indisputable, while those of all other sciences are to some extent debatable and in constant danger of being overthrown by newly discovered facts. In spite of this, the investigator in

---

* This criterion has since been confirmed.

another department of science would not need to envy the
mathematician if the propositions of mathematics referred to
objects of our mere imagination, and not to objects of reality.
For it cannot occasion surprise that different persons should
arrive at the same logical conclusions when they have already
agreed upon the fundamental propositions (axioms), as well as
the methods by which other propositions are to be deduced
therefrom. But there is another reason for the high repute of
mathematics, in that it is mathematics which affords the exact
natural sciences a certain measure of certainty, to which without mathematics they could not attain.

At this point an enigma presents itself which in all ages has
agitated inquiring minds. How can it be that mathematics, being after all a product of human thought which is independent
of experience, is so admirably appropriate to the objects of
reality? Is human reason, then, without experience, merely by
taking thought, able to fathom the properties of real things?

In my opinion the answer to this question is, briefly, this: as
far as the propositions of mathematics refer to reality, they are
not certain; and as far as they are certain, they do not refer to
reality. It seems to me that complete clarity as to this state of
things became common property only through that trend in
mathematics which is known by the name of "axiomatics." The
progress achieved by axiomatics consists in its having neatly
separated the logical-formal from its objective or intuitive content; according to axiomatics the logical-formal alone forms the
subject matter of mathematics, which is not concerned with
the intuitive or other-content associated with the logical-formal.

Let us for a moment consider from this point of view any
axiom of geometry, for instance, the following: through two
points in space there always passes one and only one straight
line. How is this axiom to be interpreted in the older sense
and in the more modern sense?

The older interpretation: everyone knows what a straight line
is, and what a point is. Whether this knowledge springs from
an ability of the human mind or from experience, from some
cooperation of the two or from some other source, is not for the

mathematician to decide. He leaves the question to the philosopher. Being based upon this knowledge, which precedes all mathematics, the axiom stated above is, like all other axioms, self-evident, that is, it is the expression of a part of this *a priori* knowledge.

The more modern interpretation: geometry treats of objects which are denoted by the words straight line, point, etc. No knowledge or intuition of these objects is assumed but only the validity of the axioms, such as the one stated above, which are to be taken in a purely formal sense, i.e., as void of all content of intuition or experience. These axioms are free creations of the human mind. All other propositions of geometry are logical inferences from the axioms (which are to be taken in the nominalistic sense only). The axioms *define* the objects of which geometry treats. Schlick in his book on epistemology has therefore characterized axioms very aptly as "implicit definitions."

This view of axioms, advocated by modern axiomatics, purges mathematics of all extraneous elements, and thus dispels the mystic obscurity which formerly surrounded the basis of mathematics. But such an expurgated exposition of mathematics makes it also evident that mathematics as such cannot predicate anything about objects of our intuition or real objects. In axiomatic geometry the words "point," "straight line," etc., stand only for empty conceptual schemata. That which gives them content is not relevant to mathematics.

Yet on the other hand it is certain that mathematics generally, and particularly geometry, owes its existence to the need which was felt of learning something about the behavior of real objects. The very word geometry, which, of course, means earth-measuring, proves this. For earth-measuring has to do with the possibilities of the disposition of certain natural objects with respect to one another, namely, with parts of the earth, measuring-lines, measuring-wands, etc. It is clear that the system of concepts of axiomatic geometry alone cannot make any assertions as to the behavior of real objects of this kind, which we will call practically-rigid bodies. To be able to make such assertions, geometry must be stripped of its merely logical-formal character by the coordination of real objects of experience with the empty conceptual schemata of axiomatic geometry. To accomplish this, we need only add the proposition: solid bodies are related, with respect to their possible dispositions, as are bodies in Euclidean geometry of three dimensions. Then the propositions of Euclid contain affirmations as to the behavior of practically-rigid bodies.

Geometry thus completed is evidently a natural science; we may in fact regard it as the most ancient branch of physics. Its affirmations rest essentially on induction from experience, but not on logical inferences only. We will call this completed geometry "practical geometry," and shall distinguish it in what follows from "purely axiomatic geometry." The question whether the practical geometry of the universe is Euclidean or not has a clear meaning, and its answer can only be furnished by experience. All length-measurements in physics constitute practical geometry in this sense, so, too, do geodetic and astronomical length measurements, if one utilizes the empirical law that light is propagated in a straight line, and indeed in a straight line in the sense of practical geometry.

I attach special importance to the view of geometry which I have just set forth, because without it I should have been unable to formulate the theory of relativity. Without it the following reflection would have been impossible: in a system of reference rotating relatively to an inertial system, the laws of disposition of rigid bodies do not correspond to the rules of Euclidean geometry on account of the Lorentz contraction; thus if we admit non-inertial systems on an equal footing, we must abandon Euclidean geometry. Without the above interpretation the decisive step in the transition to generally covariant equations would certainly not have been taken. If we reject the relation between the body of axiomatic Euclidean geometry and the practically-rigid body of reality, we readily arrive at the following view, which was entertained by that acute and profound thinker, H. Poincaré: Euclidean geometry is distinguished above all other conceivable axiomatic geometries by its simplicity. Now since axiomatic geometry by itself contains no assertions as to the reality which can be experienced, but can do so only in combination with physical laws, it should be possible and reasonable—whatever may be the nature of reality—to retain Euclidean geometry. For if contradictions between theory and experience manifest themselves, we should rather decide to change physical laws than to change axiomatic Euclidean geometry. If we reject the relation between the practically-rigid body and geometry, we shall indeed not easily free ourselves from the convention that Euclidean geometry is to be retained as the simplest.

Why is the equivalence of the practically-rigid body and the body of geometry—which suggests itself so readily—rejected by Poincaré and other investigators? Simply because under closer inspection the real solid bodies in nature are not rigid, because their geometrical behavior, that is, their possibilities of relative disposition, depend upon temperature, external forces, etc. Thus the original, immediate relation between geometry and physical reality appears destroyed, and we feel impelled toward the following more general view, which characterizes Poincaré's standpoint. Geometry (G) predicates nothing about the behavior of real things, but only geometry together with the totality (P) of physical laws can do so. Using symbols, we may say that only the sum of (G) + (P) is subject to experimental verification. Thus (G) may be chosen arbitrarily, and also parts of (P); all these laws are conventions. All that is necessary to avoid contradictions is to choose the remainder of (P) so that (G) and the whole of (P) are together in accord with experience. Envisaged in this way, axiomatic geometry and the part of natural law which has been given a conventional status appear as epistemologically equivalent.

Sub specie aeterni Poincaré, in my opinion, is right. The idea of the measuring-rod and the idea of the clock coordinated with it in the theory of relativity do not find their exact correspondence in the real world. It is also clear that the solid body and the clock do not in the conceptual edifice of physics play the part of irreducible elements, but that of composite structures, which must not play any independent part in theoretical physics. But it is my conviction that in the present stage of development of theoretical physics these concepts must still be employed as independent concepts; for we are still far from possessing such certain knowledge of the theoretical principles of atomic structure as to be able to construct solid bodies and clocks theoretically from elementary concepts.

Further, as to the objection that there are no really rigid bodies in nature, and that therefore the properties predicated of rigid bodies do not apply to physical reality—this objection is by no means so radical as might appear from a hasty examination. For it is not a difficult task to determine the physical state of a measuring-body so accurately that its behavior relative to other measuring-bodies shall be sufficiently free from ambiguity to allow it to be substituted for the "rigid" body. It is to measuring-bodies of this kind that statements about rigid bodies must be referred.

All practical geometry is based upon a principle which is accessible to experience, and which we will now try to realize. Suppose two marks have been put upon a practically-rigid body. A pair of two such marks we shall call a tract. We imagine two practically-rigid bodies, each with a tract marked out on it. These two tracts are said to be "equal to one another" if the marks of the one tract can be brought to coincide permanently with the marks of the other. We now assume that:

If two tracts are found to be equal once and anywhere, they are equal always and everywhere.

Not only the practical geometry of Euclid, but also its nearest generalization, the practical geometry of Riemann, and therewith the general theory of relativity, rest upon this assumption. Of the experimental reasons which warrant this assumption I will mention only one. The phenomenon of the propagation of light in empty space assigns a tract, namely, the appropriate path of light, to each interval of local time, and conversely. Thence it follows that the above assumption for tracts must also hold good for intervals of clock-time in the theory of relativity. Consequently it may be formulated as follows: if two ideal clocks are going at the same rate at any time and at any place (being then in immediate proximity to each other), they will always go at the same rate, no matter where and when they are again compared with each other at one place. If this law were not valid for natural clocks, the proper frequencies for the separate atoms of the same chemical element would not be in such exact agreement as experience demonstrates. The existence of sharp spectral lines is a convincing experimental proof of the above-mentioned principle of practical geometry. This, in the last analysis, is the reason which enables us to speak meaningfully of a Riemannian metric of the four-dimensional space-time continuum.

According to the view advocated here, the question whether this continuum has a Euclidean, Riemannian, or any other structure is a question of physics proper which must be answered by experience, and not a question of a convention to be chosen on grounds of mere expediency. Riemann's geometry will hold if the laws of disposition of practically-rigid bodies approach those of Euclidean geometry the more closely the smaller the dimensions of the region of space-time under consideration.

It is true that this proposed physical interpretation of geometry breaks down when applied immediately to spaces of sub-molecular order of magnitude. But nevertheless, even in questions as to the constitution of elementary particles, it retains part of its significance. For even when it is a question of describing the electrical elementary particles constituting matter, the attempt may still be made to ascribe physical meaning to those field concepts which have been physically defined for the purpose of describing the geometrical behavior of bodies which are large as compared with the molecule. Success alone can decide as to the justification of such an attempt, which postulates physical reality for the fundamental principles of Riemann's geometry outside of the domain of their physical definitions. It might possibly turn out that this extrapolation has no better warrant than the extrapolation of the concept of temperature to parts of a body of molecular order of magnitude.

It appears less problematical to extend the concepts of practical geometry to spaces of cosmic order of magnitude. It might, of course, be objected that a construction composed of solid rods departs the more from ideal rigidity the greater its spatial extent. But it will hardly be possible, I think, to assign fundamental significance to this objection. Therefore the question whether the universe is spatially finite or not seems to me an entirely meaningful question in the sense of practical geometry. I do not even consider it impossible that this question will be answered before long by astronomy. Let us call to mind what the general theory of relativity teaches in this respect. It offers two possibilities:

1. The universe is spatially infinite. This is possible only if in the universe the average spatial density of matter, concentrated in the stars, vanishes, i.e., if the ratio of the total mass of the stars to the volume of the space through which they are scattered indefinitely approaches zero as greater and greater volumes are considered.

2. The universe is spatially finite. This must be so, if there exists an average density of the ponderable matter in the universe which is different from zero. The smaller that average density, the greater is the volume of the universe.

I must not fail to mention that a theoretical argument can be adduced in favor of the hypothesis of a finite universe. The general theory of relativity teaches that the inertia of a given body is greater as there are more ponderable masses in proximity to it; thus it seems very natural to reduce the total inertia of a body to interaction between it and the other bodies in the universe, as indeed, ever since Newton's time, gravity has been completely reduced to interaction between bodies. From the equations of the general theory of relativity it can be deduced that this total reduction of inertia to interaction between masses —as demanded by E. Mach, for example—is possible only if the universe is spatially finite.

Many physicists and astronomers are not impressed by this argument. In the last analysis, experience alone can decide which of the two possibilities is realized in nature. How can experience furnish an answer? At first it might seem possible to determine the average density of matter by observation of that part of the universe which is accessible to our observation. This hope is illusory. The distribution of the visible stars is extremely irregular, so that we on no account may venture to set the average density of star-matter in the universe equal to, let us say, the average density in the Galaxy. In any case, however great the space examined may be, we could not feel convinced that there were any more stars beyond that space. So it seems impossible to estimate the average density.

But there is another road, which seems to me more practicable, although it also presents great difficulties. For if we inquire into the deviations of the consequences of the general theory of relativity which are accessible to experience, from the consequences of the Newtonian theory, we first of all find a deviation which manifests itself in close proximity to gravitating mass, and has been confirmed in the case of the planet Mercury. But if the universe is spatially finite, there is a second deviation from the Newtonian theory, which, in the language of the Newtonian theory, may be expressed thus: the gravitational field is such as if it were produced, not only by the ponderable masses, but in addition by a mass-density of negative sign, distributed uniformly throughout space. Since this fictitious mass-density would have to be extremely small, it would be noticeable only in very extensive gravitating systems.

Assuming that we know, let us say, the statistical distribution and the masses of the stars in the Galaxy, then by Newton's law we can calculate the gravitational field and the average velocities which the stars must have, so that the Galaxy should not collapse under the mutual attraction of its stars, but should maintain its actual extent. Now if the actual velocities of the stars—which can be measured—were smaller than the calculated velocities, we should have a proof that the actual attractions at great distances are smaller than by Newton's law. From such a deviation it could be proved indirectly that the universe is finite. It would even be possible to estimate its spatial dimensions.

Can we visualize a three-dimensional universe which is finite, yet unbounded?

The usual answer to this question is "No," but that is not the right answer. The purpose of the following remarks is to show that the answer should be "Yes." I want to show that without any extraordinary difficulty we can illustrate the theory of a finite universe by means of a mental picture to which, with some practice, we shall soon grow accustomed.

First of all, an observation of epistemological nature. A geometrical-physical theory as such is incapable of being directly pictured, being merely a system of concepts. But these concepts serve the purpose of bringing a multiplicity of real or imaginary sensory experiences into connection in the mind. To "visualize" a theory therefore means to bring to mind that abundance of sensible experiences for which the theory supplies the schematic arrangement. In the present case we have to ask ourselves how we can represent that behavior of solid bodies with respect to their mutual disposition (contact) which corresponds to the theory of a finite universe. There is really nothing new in what I have to say about this; but innumerable questions addressed to me prove that the curiosity of those who are interested in these matters has not yet been completely satisfied. So, will the initiated please pardon me, in that part of what I shall say has long been known?

What do we wish to express when we say that our space is infinite? Nothing more than that we might lay any number of bodies of equal sizes side by side without ever filling space. Suppose that we are provided with a great many cubic boxes all of the same size. In accordance with Euclidean geometry we can place them above, beside, and behind one another so as to fill an arbitrarily large part of space; but this construction would never be finished; we could go on adding more and more cubes without ever finding that there was no more room. That is what we wish to express when we say, that space is infinite. It would be better to say that space is infinite in relation to practically-rigid bodies, assuming that the laws of disposition for these bodies are given by Euclidean geometry.

Another example of an infinite continuum is the plane. On a plane surface we may lay squares of cardboard so that each side of any square has the side of another square adjacent to it. The construction is never finished; we can always go on laying squares—if their laws of disposition correspond to those of plane figures of Euclidean geometry. The plane is therefore infinite in relation to the cardboard squares. Accordingly we say that the plane is an infinite continuum of two dimensions, and space an infinite continuum of three dimensions. What is here meant by the number of dimensions, I think I may assume to be known.

Now we take an example of a two-dimensional continuum which is finite, but unbounded. We imagine the surface of a large globe and a quantity of small paper discs, all of the same size. We place one of the discs anywhere on the surface of the globe. If we move the disc about, anywhere we like, on the surface of the globe, we do not come upon a boundary anywhere on the journey. Therefore we say that the spherical surface of the globe is an unbounded continuum. Moreover, the spherical surface is a finite continuum. For if we stick the paper discs on the globe, so that no disc overlaps another, the surface of the globe will finally become so full that there is no room for another disc. This means exactly that the spherical surface of the globe is finite in relation to the paper discs. Further, the spherical surface is a non-Euclidean continuum of two dimensions, that is to say, the laws of disposition for the rigid figures lying in it do not agree with those of the Euclidean plane. This can be shown in the following way. Take a disc and surround it in a circle by six more discs, each of which is to be surrounded in turn by six discs, and so on. If this construction is made on a plane surface, we obtain an uninterrupted arrangement in which there are six discs touching every disc except those which lie on the outside. On the spherical surface the construction also

FIG. 1

seems to promise success at the outset, and the smaller the radius of the disc in proportion to that of the sphere, the more promising it seems. But as the construction progresses it becomes more and more patent that the arrangement of the discs in the manner indicated, without interruption, is not possible, as it should be possible by the Euclidean geometry of the plane. In this way creatures which cannot leave the spherical surface, and cannot even peep out from the spherical surface into three-dimensional space, might discover, merely by experimenting with discs, that their two-dimensional "space" is not Euclidean, but spherical space.

From the latest results of the theory of relativity it is probable that our three-dimensional space is also approximately spherical, that is, that the laws of disposition of rigid bodies in it are not given by Euclidean geometry, but approximately by spherical geometry, if only we consider parts of space which are sufficiently extended. Now this is the place where the reader's imagination boggles. "Nobody can imagine this thing," he cries indignantly. "It can be said, but cannot be thought. I can imagine a spherical surface well enough, but nothing analogous to it in three dimensions."

We must try to surmount this barrier in the mind, and the patient reader will see that it is by no means a particularly difficult task. For this purpose we will first give our attention once more to the geometry of two-dimensional spherical surfaces. In the adjoining figure let K be the spherical surface, touched at S by a plane, E, which, for facility of presentation, is shown in the drawing as a bounded surface. Let L be a disc on the spherical surface. Now let us imagine that at the point N of the

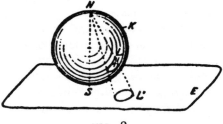

FIG. 2

spherical surface, diametrically opposite to S, there is a luminous point, throwing a shadow L' of the disc L upon the plane E. Every point on the sphere has its shadow on the plane. If the disc on the sphere K is moved, its shadow L' on the plane E also moves. When the disc L is at S, it almost exactly coincides with its shadow. If it moves on the spherical surface away from S upwards, the disc shadow L' on the plane also moves away from S on the plane outwards, growing bigger and bigger. As the disc L approaches the luminous point N, the shadow moves off to infinity, and becomes infinitely great.

Now we put the question: what are the laws of disposition of the disc-shadows L' on the plane E? Evidently they are exactly the same as the laws of disposition of the discs L on the spherical surface. For to each original figure on K there is a corresponding shadow figure on E. If two discs on K are touching, their shadows on E also touch. The shadow-geometry on the plane agrees with the disc-geometry on the sphere. If we call the disc-shadows rigid figures, then spherical geometry holds good on the plane E with respect to these rigid figures. In particular, the plane is finite with respect to the disc-shadows, since only a finite number of the shadows can find room on the plane.

At this point somebody will say, "That is nonsense. The disc-shadows are *not* rigid figures. We have only to move a two-foot rule about on the plane E to convince ourselves that the shadows constantly increase in size as they move away from S on the plane toward infinity." But what if the two-foot rule were to behave on the plane E in the same way as the disc-shadows L'? It would then be impossible to show that the shadows increase in size as they move away from S; such an assertion would then no longer have any meaning whatever. In fact the only objective assertion that can be made about the disc-shadows is just this, that they are related in exactly the same way as are the rigid discs on the spherical surface in the sense of Euclidean geometry.

We must carefully bear in mind that our statement as to the growth of the disc-shadows, as they move away from S toward infinity, has in itself no objective meaning, as long as we are unable to compare the disc-shadows with Euclidean rigid bodies which can be moved about on the plane E. In respect of the laws of disposition of the shadows L', the point S has no special privileges on the plane any more than on the spherical surface.

The representation given above of spherical geometry on the plane is important for us, because it readily allows itself to be transferred to the three-dimensional case.

Let us imagine a point S of our space, and a great number of small spheres, L', which can all be brought to coincide with one another. But these spheres are not to be rigid in the sense of Euclidean geometry; their radius is to increase (in the sense of Euclidean geometry) when they are moved away from S toward infinity; it is to increase according to the same law as the radii of the disc-shadows L' on the plane.

After having gained a vivid mental image of the geometrical behavior of our L' spheres, let us assume that in our space there are no rigid bodies at all in the sense of Euclidean geometry, but only bodies having the behavior of our L' spheres. Then we shall have a clear picture of three-dimensional spherical space, or, rather of three-dimensional spherical geometry. Here our spheres must be called "rigid" spheres. Their increase in size as they depart from S is not to be detected by measuring with measuring-rods, any more than in the case of the disc-shadows on E, because the standards of measurement behave in the same way as the spheres. Space is homogeneous, that is to say, the same spherical configurations are possible in the neighborhood of every point.* Our space is finite, because, in consequence of the "growth" of the spheres, only a finite number of them can find room in space.

In this way, by using as a crutch the practice in thinking and visualization which Euclidean geometry gives us, we have acquired a mental picture of spherical geometry. We may without difficulty impart more depth and vigor to these ideas by carrying out special imaginary constructions. Nor would it be difficult to represent the case of what is called elliptical geometry in

---

* This is intelligible without calculation—but only for the two-dimensional case—if we revert once more to the case of the disc on the surface of the sphere.

an analogous manner. My only aim today has been to show that the human faculty of visualization is by no means bound to capitulate to non-Euclidean geometry.

ON THE THEORY OF RELATIVITY

Lecture at King's College, London, 1921. Published in Mein Weltbild, *Amsterdam: Querido Verlag, 1934.*

It is a particular pleasure to me to have the privilege of speaking in the capital of the country from which the most important fundamental notions of theoretical physics have issued. I am thinking of the theory of mass motion and gravitation which Newton gave us and the concept of the electromagnetic field, by means of which Faraday and Maxwell put physics on a new basis. The theory of relativity may indeed be said to have put a sort of finishing touch to the mighty intellectual edifice of Maxwell and Lorentz, inasmuch as it seeks to extend field physics to all phenomena, gravitation included.

Turning to the theory of relativity itself, I am anxious to draw attention to the fact that this theory is not speculative in origin; it owes its invention entirely to the desire to make physical theory fit observed fact as well as possible. We have here no revolutionary act but the natural continuation of a line that can be traced through centuries. The abandonment of certain notions connected with space, time, and motion hitherto treated as fundamentals must not be regarded as arbitrary, but only as conditioned by observed facts.

The law of the constant velocity of light in empty space, which has been confirmed by the development of electrodynamics and optics, and the equal legitimacy of all inertial systems (special principle of relativity), which was proved in a particularly incisive manner by Michelson's famous experiment, between them made it necessary, to begin with, that the concept of time should be made relative, each inertial system being given its own special time. As this notion was developed, it became clear that the connection between immediate experience on one side and coordinates and time on the other had hitherto not been thought out with sufficient precision. It is in general one of the essential features of the theory of relativity that it is at pains to work out the relations between general concepts and empirical facts more precisely. The fundamental principle here is that the justification for a physical concept lies exclusively in its clear and unambiguous relation to facts that can be experienced. According to the special theory of relativity, spatial coordinates and time still have an absolute character in so far as they are directly measurable by stationary clocks and bodies. But they are relative in so far as they depend on the state of motion of the selected inertial system. According to the special theory of relativity the four-dimensional continuum formed by the union of space and time (Minkowski) retains the absolute character which, according to the earlier theory, belonged to both space and time separately. The influence of motion (relative to the coordinate system) on the form of bodies and on the motion of clocks, also the equivalence of energy and inert mass, follow from the interpretation of coordinates and time as products of measurement.

The general theory of relativity owes its existence in the first place to the empirical fact of the numerical equality of the inertial and gravitational mass of bodies, for which fundamental fact classical mechanics provided no interpretation. Such an interpretation is arrived at by an extension of the principle of relativity to coordinate systems accelerated relatively to one another. The introduction of coordinate systems accelerated relatively to inertial systems involves the appearance of gravitational fields relative to the latter. As a result of this, the general theory of relativity, which is based on the equality of inertia and weight, provides a theory of the gravitational field.

The introduction of coordinate systems accelerated relatively to each other as equally legitimate systems, such as they appear conditioned by the identity of inertia and weight, leads, in conjunction with the results of the special theory of relativity, to the conclusion that the laws governing the arrangement of solid bodies in space, when gravitational fields are present, do not correspond to the laws of Euclidean geometry. An analogous result follows for the motion of clocks. This brings us to the necessity for yet another generalization of the theory of space and time, because the direct interpretation of spatial and temporal coordinates by means of measurements obtainable with measuring rods and clocks now breaks down. That generalization of metric, which had already been accomplished in the sphere of pure mathematics through the researches of Gauss and Riemann, is essentially based on the fact that the metric of the special theory of relativity can still claim validity for small regions in the general case as well.

The process of development here sketched strips the space-time coordinates of all independent reality. The metrically real is now only given through the combination of the space-time coordinates with the mathematical quantities which describe the gravitational field.

There is yet another factor underlying the evolution of the general theory of relativity. As Ernst Mach insistently pointed out, the Newtonian theory is unsatisfactory in the following respect: if one considers motion from the purely descriptive, not from the causal, point of view, it only exists as relative motion of things with respect to one another. But the acceleration which figures in Newton's equations of motion is unintelligible if one starts with the concept of relative motion. It compelled Newton to invent a physical space in relation to which acceleration was supposed to exist. This introduction *ad hoc* of the concept of absolute space, while logically unexceptionable, nevertheless seems unsatisfactory. Hence Mach's attempt to alter the mechanical equations in such a way that the inertia of bodies is traced back to relative motion on their part not as against absolute space but as against the totality of other ponderable bodies. In the state of knowledge then existing, his attempt was bound to fail.

The posing of the problem seems, however, entirely reasonable. This line of argument imposes itself with considerably enhanced force in relation to the general theory of relativity, since, according to that theory, the physical properties of space are affected by ponderable matter. In my opinion the general theory of relativity can solve this problem satisfactorily only if it regards the world as spatially closed. The mathematical results of the theory force one to this view, if one believes that the mean density of ponderable matter in the world possesses some finite value, however small.

THE CAUSE OF THE FORMATION OF MEANDERS IN THE COURSES OF RIVERS AND OF THE SO-CALLED BAER'S LAW

Read before the Prussian Academy, January 7, 1926. Published in the German periodical, Die Naturwissenschaften, *Vol. 14, 1926.*

It is common knowledge that streams tend to curve in serpentine shapes instead of following the line of the maximum declivity of the ground. It is also well known to geographers that the rivers of the northern hemisphere tend to erode chiefly on the right side. The rivers of the southern hemisphere behave in the opposite manner (Baer's law). Many attempts have been made to explain this phenomenon, and I am not sure whether anything I say in the following pages will be new to the expert; some of my considerations are certainly known. Nevertheless, having found nobody who was thoroughly familiar with the causal relations involved, I think it is appropriate to give a short qualitative exposition of them.

First of all, it is clear that the erosion must be stronger the greater the velocity of the current where it touches the bank in question, or rather the more steeply it falls to zero at any particular point of the confining wall. This is equally true under all circumstances, whether the erosion depends on mechanical or on physico-chemical factors (decomposition of the ground). We must then concentrate our attention on the circumstances which affect the steepness of the velocity gradient at the wall.

In both cases the asymmetry as regards the fall in velocity in question is indirectly due to the formation of a circular motion to which we will next direct our attention.

I begin with a little experiment which anybody can easily repeat. Imagine a flat-bottomed cup full of tea. At the bottom there are some tea leaves, which stay there because they are rather heavier than the liquid they have displaced. If the liquid is made to rotate by a spoon, the leaves will soon collect in the center of the bottom of the cup. The explanation of this phenomenon is as follows: the rotation of the liquid causes a centrifugal force to act on it. This in itself would give rise to no change in the flow of the liquid if the latter rotated like a solid body. But in the neighborhood of the walls of the cup the liquid is restrained by friction, so that the angular velocity with which it rotates is less there than in other places nearer the

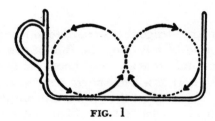

FIG. 1

center. In particular, the angular velocity of rotation, and therefore the centrifugal force, will be smaller near the bottom than higher up. The result of this will be a circular movement of the liquid of the type illustrated in Fig. 1 which goes on increasing until, under the influence of ground friction, it becomes stationary. The tea leaves are swept into the center by the circular movement and act as proof of its existence.

The same sort of thing happens with a curving stream (Fig. 2). At every cross-section of its course, where it is bent, a centrifugal force operates in the direction of the outside of the curve (from *A* to *B*). This force is less near the bottom, where the speed of the current is reduced by friction, than higher above the bottom. This causes a circular movement of the kind illustrated in the diagram. Even where there is no bend in the river, a circular movement of the kind shown in Fig. 2 will still take place, if only on a small scale, as a result of the earth's rotation. The

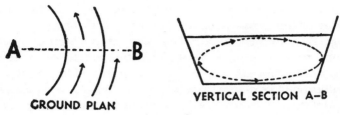

GROUND PLAN   VERTICAL SECTION A–B

FIG. 2

latter produces a Coriolis-force, acting transversely to the direction of the current, whose right-hand horizontal component amounts to $2 \, v \, \Omega \sin \phi$ per unit of mass of the liquid, where v is the velocity of the current, Ω the speed of the earth's rotation, and ϕ the geographical latitude. As ground friction causes a diminution of this force toward the bottom, this force also gives rise to a circular movement of the type indicated in Fig. 2.

After this preliminary discussion we come back to the question of the distribution of velocities over the cross-section of the stream, which is the controlling factor in erosion. For this purpose we must first realize how the (turbulent) distribution of velocities develops and is maintained. If the water which was previously at rest were suddenly set in motion by the action of a uniformly distributed accelerating force, the distribution of velocities over the cross-section would at first be uniform. A distribution of velocities gradually increasing from the confining walls toward the center of the cross-section would only establish itself after a time, under the influence of friction at the walls. A disturbance of the (roughly speaking) stationary distribution of velocities over the cross-section would only gradually set in again under the influence of fluid friction.

Hydrodynamics pictures the process by which this stationary distribution of velocities is established in the following way. In a plane (potential) flow all the vortex-filaments are concentrated at the walls. They detach themselves and slowly move toward the center of the cross-section of the stream, distributing themselves over a layer of increasing thickness. The velocity gradient at the walls thereby gradually diminishes. Under the action of the internal friction of the liquid the vortex filaments in the interior of the cross-section are gradually absorbed, their place being taken by new ones which form at the wall. A quasi-stationary distribution of velocities is thus produced. The important thing for us is that the attainment of the stationary distribution of velocities is a slow process. That is why relatively insignificant, constantly operative causes are able to exert a considerable influence on the distribution of velocities over the cross-section. Let us now consider what sort of influence the circular motion due to a bend in the river or the Coriolis-force, as illustrated in Fig. 2, is bound to exert on the distribution of velocities over the cross-section of the river. The particles of liquid in most rapid motion will be farthest away from the walls, that is to say, in the upper part above the center of the bottom. These most rapid parts of the water will be driven by the circulation toward the right-hand wall, while the left-hand wall gets the water which comes from the region near the bottom and has a specially low velocity. Hence in the case depicted in Fig. 2 the erosion is necessarily stronger on the right side than on the left. It should be noted that this explanation is essentially based on the fact that the slow circulating movement of the water exerts a considerable influence on the distribution of velocities, because the adjustment of velocities by internal friction which counteracts this consequence of the circulating movement is also a slow process.

We have now revealed the causes of the formation of meanders. Certain details can, however, also be deduced without difficulty from these facts. Erosion will be comparatively extensive not merely on the right-hand wall but also on the right half of the bottom, so that there will be a tendency to assume a profile as illustrated in Fig. 3.

Moreover, the water at the surface will come from the left-hand wall, and will therefore, on the left-hand side especially,

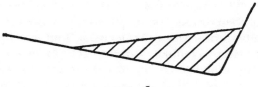

FIG. 3

be moving less rapidly than the water rather lower down. This has, in fact, been observed. It should further be noted that the circular motion possesses inertia. The circulation will therefore only achieve its maximum beyond the place of the greatest curvature, and the same naturally applies to the asymmetry of the erosion. Hence in the course of the erosion an advance of the wave-line of the meander-formation is bound to take place in the direction of the current. Finally, the larger the cross-section of the river, the more slowly will the circular movement be absorbed by friction; the wave-line of the meander-formation will therefore increase with the cross-section of the river.

THE MECHANICS OF NEWTON AND THEIR INFLUENCE ON THE DEVELOPMENT OF THEORETICAL PHYSICS

On the occasion of the two hundreth anniversary of Newton's death. Published in Vol. 15 of the German periodical, Die Naturwissenschaften, *1927.*

It is just two hundred years ago that Newton closed his eyes. We feel impelled at such a moment to remember this brilliant genius, who determined the course of western thought, research, and practice like no one else before or since. Not only was he brilliant as an inventor of certain key methods, but he also had a unique command of the empirical material available in his day, and he was marvelously inventive as regards detailed mathematical and physical methods of proof. For all these reasons he deserves our deepest reverence. The figure of Newton has, however, an even greater importance than his genius warrants because destiny placed him at a turning point in the history of the human intellect. To see this vividly, we have to realize that before Newton there existed no self-contained system of physical causality which was somehow capable of representing any of the deeper features of the empirical world.

No doubt the great materialists of ancient Greece had insisted that all material events should be traced back to a strictly regular series of atomic movements, without admitting any living creature's will as an independent cause. And no doubt Descartes had in his own way taken up this quest again. But it remained a bold ambition, the problematical ideal of a school of philosophers. Actual results of a kind to support the belief in the existence of a complete chain of physical causation hardly existed before Newton.

Newton's object was to answer the question: is there any simple rule by which one can calculate the movements of the heavenly bodies in our planetary system completely, when the state of motion of all these bodies at one moment is known? Kepler's empirical laws of planetary movement, deduced from Tycho Brahe's observations, confronted him, and demanded explanation.* These laws gave, it is true, a complete answer to the question of *how* the planets move round the sun: the elliptical shape of the orbit, the sweeping of equal areas by the radii in equal times, the relation between the major axes and the periods of revolution. But these rules do not satisfy the demand for causal explanation. They are three logically independent rules, revealing no inner connection with each other. The third law cannot simply be transferred quantitatively to other central bodies than the sun (there is, e.g., no relation between the period of revolution of a planet round the sun and

---

* Today everybody knows what prodigious industry was needed to discover these laws from the empirically ascertained orbits. But few pause to reflect on the brilliant method by which Kepler deduced the real orbits from the apparent ones—i.e., from the movements as they were observed from the earth.

that of a moon round its planet). The most important point, however, is this: these laws are concerned with the movement as a whole, and not with the question *how the state of motion of a system gives rise to that which immediately follows it in time;* they are, as we should say now, integral and not differential laws.

The differential law is the only form which completely satisfies the modern physicist's demand for causality. The clear conception of the differential law is one of Newton's greatest intellectual achievements. It was not merely this conception that was needed but also a mathematical formalism, which existed in a rudimentary way but needed to acquire a systematic form. Newton found this also in the differential and the integral calculus. We need not consider the question here whether Leibnitz hit upon the same mathematical methods independently of Newton, or not. In any case it was absolutely necessary for Newton to perfect them, since they alone could provide him with the means of expressing his ideas.

Galileo had already made a significant beginning toward a knowledge of the law of motion. He discovered the law of inertia and the law of bodies falling freely in the gravitational field of the earth, namely, that a mass (more accurately, a mass-point) which is unaffected by other masses moves uniformly and in a straight line. The vertical speed of a free body in the gravitational field increases uniformly with time. It may seem to us today to be but a short step from Galileo's discoveries to Newton's law of motion. But it should be observed that both the above statements are so formulated as to refer to the motion as a whole, while Newton's law of motion provides an answer to the question: how does the state of motion of a mass-point change in an infinitely short time under the influence of an external force? It was only by considering what takes place during an infinitely short time (differential law) that Newton reached a formulation which applies to all motion whatsoever. He took the concept of force from the science of statics which had already reached a high stage of development. He was only able to connect force and acceleration by introducing the new concept of mass, which was supported, strange to say, by an illusory definition. We are so accustomed today to form concepts corresponding to differential quotients that we can now hardly grasp any longer what a remarkable power of abstraction it needed to obtain the general differential law by a double limiting process in the course of which the concept of mass had in addition to be invented.

But a causal concept of motion was still far from being achieved. For the motion was only determined by the equation of motion in cases where the force was given. Inspired no doubt by the laws of planetary motions, Newton conceived the idea that the force operating on a mass was determined by the position of all masses situated at a sufficiently small distance from the mass in question. It was not till this connection was established that a completely causal concept of motion was achieved. How Newton, starting from Kepler's laws of planetary motion, performed this task for gravitation and so discovered that the moving forces acting on the stars and gravity were of the same nature, is well known. It is the combination

Law of Motion plus Law of Attraction

which constitutes that marvelous edifice of thought which makes it possible to calculate the past and future states of a system from the state obtaining at one particular moment, in so far as the events take place under the influence of the forces of gravity alone. The logical completeness of Newton's conceptual system lay in this, that the only causes of the acceleration of the masses of a system are *these masses themselves*.

On the basis of the foundation here briefly sketched, Newton succeeded in explaining the motions of the planets, moons, and comets down to the smallest details, as well as the tides and the precessional movement of the earth—a deductive achievement of unique magnificence. The discovery that the cause of the motions of the heavenly bodies is identical with the gravity with which we are so familiar from everyday life must have been particularly impressive.

But the importance of Newton's achievement was not confined to the fact that it created a workable and logically satisfactory basis for the actual science of mechanics; up to the end of the nineteenth century it formed the program of every worker in the field of theoretical physics. All physical events were to be traced back to masses subject to Newton's laws of motion. The law of force simply had to be extended and adapted to the type of event under consideration. Newton himself tried to apply this program to optics, assuming light to consist of inert corpuscles. Even the wave theory of light made use of Newton's law of motion, after it had been applied to continuously distributed masses. Newton's equations of motion were the sole basis of the kinetic theory of heat, which not only prepared people's minds for the discovery of the law of the conservation of energy but also led to a theory of gases which has been confirmed down to the last detail, and a more profound view of the nature of the second law of thermodynamics. The development of electricity and magnetism has proceeded up to modern times along Newtonian lines (electrical and magnetic substance, forces acting at a distance). Even the revolution in electrodynamics and optics brought about by Faraday and Maxwell, which formed the first great fundamental advance in theoretical physics since Newton, took place entirely under the ægis of Newton's ideas. Maxwell, Boltzmann, and Lord Kelvin never wearied of tracing the electromagnetic fields and their dynamic interactions back to the mechanical action of hypothetical continuously distributed masses. As a result, however, of the lack of success, or at any rate of any marked success of those efforts, a gradual shift in our fundamental notions has taken place since the end of the nineteenth century; theoretical physics has outgrown the Newtonian frame which gave stability and intellectual guidance to science for nearly two hundred years.

Newton's fundamental principles were so satisfactory from the logical point of view that the impetus to overhaul them could only spring from the demands of empirical fact. Before I go into this I must emphasize that Newton himself was better aware of the weaknesses inherent in his intellectual edifice than the generations of learned scientists which followed him. This fact has always aroused my deep admiration, and I should like, therefore, to dwell on it for a moment.

I. Newton's endeavors to represent his system as necessarily conditioned by experience and to introduce the smallest possible number of concepts not directly referable to empirical objects is everywhere evident; in spite of this he set up the concept of absolute space and absolute time. For this he has often been criticized in recent years. But in this point Newton is particularly consistent. He had realized that observable geometrical quantities (distances of material points from one another) and their course in time do not completely characterize motion in its physical aspects. He proved this in the famous experiment with the rotating vessel of water. Therefore, in addition to masses and temporally variable distances, there must be something else that determines motion. That "something" he takes to be relation to "absolute space." He is aware that space must possess a kind of physical reality if his laws of motion are to have any meaning, a reality of the same sort as material points and their distances.

The clear realization of this reveals both Newton's wisdom and also a weak side to his theory. For the logical structure of the latter would undoubtedly be more satisfactory without this shadowy concept; in that case only things whose relations to perception are perfectly clear (mass-points, distances) would enter into the laws.

II. Forces acting directly and instantaneously at a distance, as introduced to represent the effects of gravity, are not in character with most of the processes familiar to us from everyday life. Newton meets this objection by pointing to the fact that his law of gravitational interaction is not supposed to be a final explanation but a rule derived by induction from experience.

III. Newton's theory provided no explanation for the highly remarkable fact that the weight and the inertia of a body are determined by the same quantity (its mass). Newton himself was aware of the peculiarity of this fact.

None of these three points can rank as a logical objection

to the theory. In a sense they merely represent unsatisfied desires of the scientific mind in its struggle for a complete and uniform conceptual grasp of natural phenomena.

Newton's theory of motion, considered as a program for the whole of theoretical physics, received its first blow from Maxwell's theory of electricity. It became clear that the electric and magnetic interactions between bodies were effected, not by forces operating instantaneously at a distance, but by processes which are propagated through space at a finite speed. In addition to the mass point and its motion, there arose according to Faraday's concept a new kind of physical reality, namely, the "field." At first people tried, adhering to the point of view of mechanics, to interpret the field as a mechanical state (of motic n or stress) of a hypothetical medium (the ether) permeating space. But when this interpretation refused to work in spite of the most obstinate efforts, people gradually got used to the idea of regarding the "electromagnetic field" as the final irreducible constituent of physical reality. We have H. Hertz to thank for definitely freeing the concept of the field from all encumbrances derived from the conceptual armory of mechanics, and H. A. Lorentz for freeing it from a material substratum; according to the latter the only thing left as substratum for the field was physical empty space (or ether), which even in the mechanics of Newton had not been destitute of all physical functions. By the time this point was reached, nobody any longer believed in immediate momentary action at a distance, not even in the sphere of gravitation, although no field theory of the latter was clearly indicated owing to lack of sufficient factual knowledge. The development of the theory of the electromagnetic field—once Newton's hypothesis of forces acting at a distance had been abandoned—led also to the attempt to explain the Newtonian law of motion on electromagnetic lines or to replace it by a more accurate one based on the field-theory. Even though these efforts did not meet with complete success, still the fundamental concepts of mechanics had ceased to be looked upon as fundamental constituents of the physical cosmos.

The theory of Maxwell and Lorentz led inevitably to the special theory of relativity, which, since it abandoned the notion of absolute simultaneity, excluded the existence of forces acting at a distance. It followed from this theory that mass is not a constant quantity but depends on (indeed it is equivalent to) the energy content. It also showed that Newton's law of motion was only to be regarded as a limiting law valid for small velocities; in its place it set up a new law of motion in which the speed of light in vacuo figures as the limiting velocity.

The general theory of relativity formed the last step in the development of the program of the field-theory. Quantitatively it modified Newton's theory only slightly, but for that all the more profoundly qualitatively. Inertia, **gravitation**, and the metrical behavior of bodies and clocks were reduced to a single field quality; this field itself was again postulated as dependent on bodies (generalization of Newton's law of gravity or rather the field law corresponding to it, as formulated by Poisson). Space and time were thereby divested not of their reality but of their causal absoluteness—i.e., affecting but not affected—which Newton had been compelled to ascribe to them in order to formulate the laws then known. The generalized law of inertia takes over the function of Newton's law of motion. This short account is enough to show how the elements of Newtonian theory passed over into the general theory of relativity, whereby the three defects above mentioned were overcome. It looks as if in the framework of the theory of general relativity the law of motion could be deduced from the field law corresponding to the Newtonian law of force. Only when this goal has been completely reached will it be possible to talk about a pure field-theory.

In a more formal sense also Newton's mechanics prepared the way for the field-theory. The application of Newton's mechanics to continuously distributed masses led inevitably to the discovery and application of partial differential equations, which in their turn first provided the language for the laws of the field-theory. In this formal respect Newton's conception of the differential law constitutes the first decisive step in the development which followed.

The whole evolution of our ideas about the processes of nature, with which we have been concerned so far, might be regarded as an organic development of Newton's ideas. But while the process of perfecting the field-theory was still in full swing, the facts of heat-radiation, the spectra, radioactivity, etc., revealed a limitation of the applicability of this whole conceptual system which today still seems to us virtually impossible to overcome notwithstanding immense successes in many instances. Many physicists maintain—and there are weighty arguments in their favor—that in the face of these facts not merely the differential law but the law of causation itself—hitherto the ultimate basic postulate of all natural science—has collapsed. Even the possibility of a spatio-temporal construction, which can be unambiguously coordinated with physical events, is denied. That a mechanical system can have only discrete permanent energy-values or states—as experience almost directly shows—seems at first sight hardly deducible from a field-theory which operates with differential equations. The de Broglie-Schrödinger method, which has in a certain sense the character of a field-theory, does indeed deduce the existence of only discrete states, in surprising agreement with empirical facts. It does so on the basis of differential equations applying a kind of resonance-argument, but it has to give up the localization of particles and strictly causal laws. Who would presume today to decide the question whether the law of causation and the differential law, these ultimate premises of the Newtonian view of nature, must definitely be abandoned?

ON SCIENTIFIC TRUTH

Answers to questions of a Japanese scholar. Published in Gelegentliches, *1929, which appeared in a limited edition on the occasion of Einstein's fiftieth birthday.*

I. It is difficult even to attach a precise meaning to the term "scientific truth." Thus the meaning of the word "truth" varies according to whether we deal with a fact of experience, a mathematical proposition, or a scientific theory. "Religious truth" conveys nothing clear to me at all.

II. Scientific research can reduce superstition by encouraging people to think and view things in terms of cause and effect. Certain it is that a conviction, akin to religious feeling, of the rationality or intelligibility of the world lies behind all scientific work of a higher order.

III. This firm belief, a belief bound up with deep feeling, in a superior mind that reveals itself in the world of experience, represents my conception of God. In common parlance this may be described as "pantheistic" (Spinoza).

IV. Denominational traditions I can only consider historically and psychologically; they have no other significance for me.

JOHANNES KEPLER

On the occasion of the three hundredth anniversary of Kepler's death. Published in the Frankfurter Zeitung *(Germany), November 9, 1930.*

In anxious and uncertain times like ours, when it is difficult to find pleasure in humanity and the course of human affairs, it is particularly consoling to think of such a supreme and quiet man as Kepler. Kepler lived in an age in which the reign of law in nature was as yet by no means certain. How great must his faith in the existence of natural law have been to give him the strength to devote decades of hard and patient work to the empirical investigation of planetary motion and the mathematical laws of that motion, entirely on his own, supported by no one and understood by very few! If we would honor his memory fittingly, we must get as clear a picture as we can of his problem and the stages of its solution.

Copernicus had opened the eyes of the most intelligent to the fact that the best way to get a clear grasp of the apparent movements of the planets in the heavens was to regard them as movements round the sun conceived as stationary. If the planets moved uniformly in a circle round the sun, it would have been comparatively easy to discover how these movements must look from the earth. Since, however, the phenomena to be dealt with were much more complicated than that, the task was far harder. First of all, these movements had to be determined empirically from the observations of Tycho Brahe. Only then did it become possible to think about discovering the general laws which these movements satisfy.

To grasp how difficult a business it was even to determine the actual movements round the sun one has to realize the following. One can never see where a planet really is at any given moment, but only in what direction it can be seen just then from the earth, which is itself moving in an unknown manner round the sun. The difficulties thus seemed practically insurmountable.

Kepler had to discover a way of bringing order into this chaos. To start with, he saw that it was necessary first to try to find out about the motion of the earth itself. This would simply have been impossible if there existed only the sun, the earth, and the fixed stars, but no other planets. For in that case one could ascertain nothing empirically except how the direction of the straight sun-earth line changes in the course of the year (apparent movement of the sun with reference to the fixed stars). In this way it was possible to discover that these sun-earth directions all lay in a plane stationary with reference to the fixed stars, at least according to the accuracy of observation achieved in those days, when there were no telescopes. By this means it could also be ascertained in what manner the line sun-earth revolves round the sun. It turned out that the angular velocity of this motion varied in a regular way in the course of the year. But this was not of much use, as it was still not known how the distance from the earth to the sun alters in the course of the year. Only when these changes were known, could the real shape of the earth's orbit and the manner in which it is described be ascertained.

Kepler found a marvelous way out of this dilemma. To begin with it followed from observations of the sun that the apparent path of the sun against the background of the fixed stars differed in speed at different times of the year, but that the angular velocity of this movement was always the same at the same time of the astronomical year, and therefore that the speed of rotation of the straight line earth-sun was always the same when it pointed to the same region of the fixed stars. It was thus legitimate to suppose that the earth's orbit was closed, described by the earth in the same way every year—which was by no means obvious *a priori*. For the adherents of the Copernican system it was thus as good as certain that this must also apply to the orbits of the rest of the planets.

This certainly made things easier. But how to ascertain the real shape of the earth's orbit? Imagine a brightly shining lantern *M* somewhere in the plane of the orbit. Assume we know that this lantern remains permanently in its place and thus forms a kind of fixed triangulation point for determining the earth's orbit, a point which the inhabitants of the earth can take a sight on at any time of year. Let this lantern *M* be further away from the sun than the earth. With the help of such a lantern it was possible to determine the earth's orbit, in the following way:

First of all, in every year there comes a moment when the earth *E* lies exactly on the line joining the sun *S* and the lantern *M*. If at this moment we look from the earth *E* at the lantern *M*, our line of sight will coincide with the line *SM* (sun-lantern). Suppose the latter to be marked in the heavens. Now imagine the earth in a different position and at a different time. Since the sun *S* and the lantern *M* can both be seen from the earth, the angle at *E* in the triangle *SEM* is known. But we also know the direction of *SE* in relation to the fixed stars through direct solar observations, while the direction of the line *SM* in relation to the fixed stars has previously been ascertained once for all. In the triangle *SEM* we also know the angle at *S*. Therefore, with the base *SM* arbitrarily laid down on a sheet of paper, we can, in virtue of our knowledge of the angles at *E* and *S*, construct the triangle *SEM*. We might do this at frequent intervals during the year; each time we should get on our piece of paper a position of the earth *E* with a date attached to it and a certain position in relation to the permanently fixed base *SM*. The earth's orbit would thereby be empirically determined, apart from its absolute size, of course.

But, you will say, where did Kepler get his lantern *M*? His genius and nature, benevolent in this case, gave it to him.

There was, for example, the planet Mars; and the length of the Martian year—i.e., one rotation of Mars round the sun—was known. At one point, it may happen that the sun, the earth, and Mars lie very nearly on a straight line. This position of Mars regularly recurs after one, two, etc., Martian years, as Mars moves in a closed orbit. At these known moments, therefore, *SM* always presents the same base, while the earth is always at a different point in its orbit. The observations of the sun and Mars at these moments thus constitute a means of determining the true orbit of the earth, as Mars then plays the part of our imaginary lantern. Thus it was that Kepler discovered the true shape of the earth's orbit and the way in which the earth describes it, and we who come after—Europeans, Germans, or even Swabians—may well admire and honor him for it.

Now that the earth's orbit had been empirically determined, the true position and length of the line *SE* at any moment was known, and it was not so terribly difficult for Kepler to calculate the orbits and motions of the rest of the planets, too, from observations—at least in principle. It was nevertheless an immense task, especially considering the state of mathematics at the time.

Now came the second and no less arduous part of Kepler's life-work. The orbits were empirically known, but their laws had to be guessed from the empirical data. First he had to make a guess at the mathematical nature of the curve described by the orbit, and then try it out on a vast assemblage of figures. If it did not fit, another hypothesis had to be devised and again tested. After tremendous search, the conjecture that the orbit was an ellipse with the sun at one of its foci was found to fit the facts. Kepler also discovered the law governing the variation in speed during one revolution, which is that the line sun-planet sweeps out equal areas in equal periods of time. Finally he also discovered that the squares of the periods of revolution round the sun vary as the cubes of the major axes of the ellipses.

Our admiration for this splendid man is accompanied by another feeling of admiration and reverence, the object of which is no man but the mysterious harmony of nature into which we are born. The ancients already devised the lines exhibiting the simplest conceivable form of regularity. Among these, next to the straight line and the circle, the most important were the ellipse and the hyperbola. We see the last two embodied—at least very nearly so—in the orbits of the heavenly bodies.

It seems that the human mind has first to construct forms independently before we can find them in things. Kepler's marvelous achievement is a particularly fine example of the truth that knowledge cannot spring from experience alone but only from the comparison of the inventions of the intellect with observed fact.

MAXWELL'S INFLUENCE ON THE EVOLUTION OF THE IDEA OF PHYSICAL REALITY

On the one hundredth anniversary of Maxwell's birth. Published, 1931, in James Clerk Maxwell: A Commemoration Volume, *Cambridge University Press.*

The belief in an external world independent of the perceiving subject is the basis of all natural science. Since, however, sense perception only gives information of this external world or of "physical reality" indirectly, we can only grasp the latter by speculative means. It follows from this that our notions of physical reality can never be final. We must always be ready to change these notions—that is to say, the axiomatic basis of physics—in order to do justice to perceived facts in the most perfect way logically. Actually a glance at the development of physics shows that it has undergone far-reaching changes in the course of time.

The greatest change in the axiomatic basis of physics—in other words, of our conception of the structure of reality—since Newton laid the foundation of theoretical physics was brought about by Faraday's and Maxwell's work on electromagnetic phenomena. We will try in what follows to make this clearer, keeping both earlier and later developments in sight.

According to Newton's system, physical reality is characterized by the concepts of space, time, material point, and force (reciprocal action of material points). Physical events, in Newton's view, are to be regarded as the motions, governed by fixed laws, of material points in space. The material point is our only mode of representing reality when dealing with changes taking place in it, the solitary representative of the real, in so far as the real is capable of change. Perceptible bodies are obviously responsible for the concept of the material point; people conceived it as an analogue of mobile bodies, stripping these of the characteristics of extension, form, orientation in space, and all "inward" qualities, leaving only inertia and translation and adding the concept of force. The material bodies, which had led psychologically to our formation of the concept of the "material point," had now themselves to be regarded as systems of material points. It should be noted that this theoretical scheme is in essence an atomistic and mechanistic one. All happenings were to be interpreted purely mechanically—that is to say, simply as motions of material points according to Newton's law of motion.

The most unsatisfactory side of this system (apart from the difficulties involved in the concept of "absolute space" which have been raised once more quite recently) lay in its description of light, which Newton also conceived, in accordance with his system, as composed of material points. Even at that time the question, What in that case becomes of the material points of which light is composed, when the light is absorbed?, was already a burning one. Moreover, it is unsatisfactory in any case to introduce into the discussion material points of quite a different sort, which had to be postulated for the purpose of representing ponderable matter and light respectively. Later on, electrical corpuscles were added to these, making a third kind, again with completely different characteristics. It was, further, a fundamental weakness that the forces of reciprocal action, by which events are determined, had to be assumed hypothetically in a perfectly arbitrary way. Yet this conception of the real accomplished much: how came it that people felt themselves impelled to forsake it?

In order to put his system into mathematical form at all, Newton had to devise the concept of differential quotients and propound the laws of motion in the form of total differential equations—perhaps the greatest advance in thought that a single individual was ever privileged to make. Partial differential equations were not necessary for this purpose, nor did Newton make any systematic use of them; but they were necessary for the formulation of the mechanics of deformable bodies; this is connected with the fact that in these problems the question of *how* bodies are supposed to be constructed out of material points was of no importance to begin with.

Thus the partial differential equation entered theoretical physics as a handmaid, but has gradually become mistress. This began in the nineteenth century when the wave-theory of light established itself under the pressure of observed fact. Light in empty space was explained as a matter of vibrations of the ether, and it seemed idle at that stage, of course, to look upon the latter as a conglomeration of material points. Here for the first time the partial differential equation appeared as the natural expression of the primary realities of physics. In a particular department of theoretical physics the continuous field thus appeared side by side with the material point as the representative of physical reality. This dualism remains even today, disturbing as it must be to every orderly mind.

If the idea of physical reality had ceased to be purely atomic, it still remained for the time being purely *mechanistic;* people still tried to explain all events as the motion of inert masses; indeed no other way of looking at things seemed conceivable. Then came the great change, which will be associated for all time with the names of Faraday, Maxwell, and Hertz. The lion's share in this revolution fell to Maxwell. He showed that the whole of what was then known about light and electromagnetic phenomena was expressed in his well-known double system of differential equations, in which the electric and the magnetic fields appear as the dependent variables. Maxwell did, indeed, try to explain, or justify, these equations by the intellectual construction of a mechanical model.

But he made use of several such constructions at the same time and took none of them really seriously, so that the equations alone appeared as the essential thing and the field strengths as the ultimate entities, not to be reduced to anything else. By the turn of the century the conception of the electromagnetic field as an ultimate entity had been generally accepted and serious thinkers had abandoned the belief in the justification, or the possibility, of a mechanical explanation of Maxwell's equations. Before long they were, on the contrary, actually trying to explain material points and their inertia on field theory lines with the help of Maxwell's theory, an attempt which did not, however, meet with complete success.

Neglecting the important *individual* results which Maxwell's life-work produced in important departments of physics, and concentrating on the changes wrought by him in our conception of the nature of physical reality, we may say this: before Maxwell people conceived of physical reality—in so far as it is supposed to represent events in nature—as material points, whose changes consist exclusively of motions, which are subject to total differential equations. After Maxwell they conceived physical reality as represented by continuous fields, not mechanically explicable, which are subject to partial differential equations. This change in the conception of reality is the most profound and fruitful one that has come to physics since Newton; but it has at the same time to be admitted that the program has by no means been completely carried out yet. The successful systems of physics which have been evolved since rather represent compromises between these two schemes, which for that very reason bear a provisional, logically incomplete character, although they may have achieved great advances in certain particulars.

The first of these that calls for mention is Lorentz's theory of electrons, in which the field and the electrical corpuscles appear side by side as elements of equal value for the comprehension of reality. Next come the special and general theories of relativity, which, though based entirely on ideas connected with the field-theory, have so far been unable to avoid the independent introduction of material points and total differential equations.

The last and most successful creation of theoretical physics, namely quantum-mechanics, differs fundamentally from both the schemes which we will for the sake of brevity call the Newtonian and the Maxwellian. For the quantities which figure in its laws make no claim to describe physical reality itself, but only the *probabilities* of the occurrence of a physical reality that we have in view. Dirac, to whom, in my opinion, we owe the most perfect exposition, logically, of this theory, rightly points out that it would probably be difficult, for example, to give a theoretical description of a photon such as would give enough information to enable one to decide whether it will pass a polarizer placed (obliquely) in its way or not.

I am still inclined to the view that physicists will not in the long run content themselves with that sort of indirect description of the real, even if the theory can eventually be adapted to the postulate of general relativity in a satisfactory manner. We shall then, I feel sure, have to return to the attempt to carry out the program which may be described properly as the Maxwellian—namely, the description of physical reality in terms of fields which satisfy partial differential equations without singularities.

ON THE METHOD OF THEORETICAL PHYSICS

The Herbert Spencer lecture, delivered at Oxford, June 10, 1933. Published in Mein Weltbild, *Amsterdam: Querido Verlag, 1934.*

If you want to find out anything from the theoretical physicists about the methods they use, I advise you to stick closely to one principle: don't listen to their words, fix your attention on their deeds. To him who is a discoverer in this field, the products of his imagination appear so necessary and natural that he regards them, and would like to have them regarded by others, not as creations of thought but as given realities.

These words sound like an invitation to you to walk out of this lecture. You will say to yourselves, the fellow's a working physicist himself and ought therefore to leave all questions of the structure of theoretical science to the epistemologists.

Against such criticism I can defend myself from the personal point of view by assuring you that it is not at my own instance but at the kind invitation of others that I have mounted this rostrum, which serves to commemorate a man who fought hard all his life for the unity of knowledge. Objectively, however, my enterprise can be justified on the ground that it may, after all, be of interest to know how one who has spent a lifetime in striving with all his might to clear up and rectify its fundamentals looks upon his own branch of science. The way in which he regards its past and present may depend too much on what he hopes for the future and aims at in the present; but that is the inevitable fate of anybody who has occupied himself intensively with a world of ideas. The same thing happens to him as to the historian, who in the same way, even though perhaps unconsciously, groups actual events round ideals which he has formed for himself on the subject of human society.

Let us now cast an eye over the development of the theoretical system, paying special attention to the relations between the content of the theory and the totality of empirical fact. We are concerned with the eternal antithesis between the two inseparable components of our knowledge, the empirical and the rational, in our department.

We reverence ancient Greece as the cradle of western science Here for the first time the world witnessed the miracle of a logi cal system which proceeded from step to step with such precision that every single one of its propositions was absolutely indubitable—I refer to Euclid's geometry. This admirable triumph of reasoning gave the human intellect the necessary confidence in itself for its subsequent achievements. If Euclid failed to kindle your youthful enthusiasm, then you were not born to be a scientific thinker.

But before mankind could be ripe for a science which takes in the whole of reality, a second fundamental truth was needed, which only became common property among philosophers with the advent of Kepler and Galileo. Pure logical thinking cannot yield us any knowledge of the empirical world; all knowledge of reality starts from experience and ends in it. Propositions arrived at by purely logical means are completely empty as regards reality. Because Galileo saw this, and particularly because he drummed it into the scientific world, he is the father of modern physics—indeed, of modern science altogether.

If, then, experience is the alpha and the omega of all our knowledge of reality, what is the function of pure reason in science?

A complete system of theoretical physics is made up of concepts, fundamental laws which are supposed to be valid for those concepts and conclusions to be reached by logical deduction. It is these conclusions which must correspond with our separate experiences; in any theoretical treatise their logical deduction occupies almost the whole book.

This is exactly what happens in Euclid's geometry, except that there the fundamental laws are called axioms and there is no question of the conclusions having to correspond to any sort of experience. If, however, one regards Euclidean geometry as the science of the possible mutual relations of practically rigid bodies in space, that is to say, treats it as a physical science, without abstracting from its original empirical content, the logical homogeneity of geometry and theoretical physics becomes complete.

We have thus assigned to pure reason and experience their places in a theoretical system of physics. The structure of the system is the work of reason; the empirical contents and their mutual relations must find their representation in the conclusions of the theory. In the possibility of such a representation lie the sole value and justification of the whole system, and especially of the concepts and fundamental principles which underlie it. Apart from that, these latter are free inventions of the human intellect, which cannot be justified either by the nature of that intellect or in any other fashion *a priori*.

These fundamental concepts and postulates, which cannot be further reduced logically, form the essential part of a theory, which reason cannot touch. It is the grand object of all theory to make these irreducible elements as simple and as few in number as possible, without having to renounce the adequate representation of any empirical content whatever.

The view I have just outlined of the purely fictitious character of the fundamentals of scientific theory was by no means the prevailing one in the eighteenth and nineteenth centuries. But it is steadily gaining ground from the fact that the distance in thought between the fundamental concepts and laws on one side and, on the other, the conclusions which have to be brought into relation with our experience grows larger and larger, the simpler the logical structure becomes—that is to say, the smaller the number of logically independent conceptual elements which are found necessary to support the structure.

Newton, the first creator of a comprehensive, workable system of theoretical physics, still believed that the basic concepts and laws of his system could be derived from experience. This is no doubt the meaning of his saying, *hypotheses non fingo*.

Actually the concepts of time and space appeared at that time to present no difficulties. The concepts of mass, inertia, and force, and the laws connecting them, seemed to be drawn directly from experience. Once this basis is accepted, the expression for the force of gravitation appears derivable from experience, and it was reasonable to expect the same in regard to other forces.

We can indeed see from Newton's formulation of it that the concept of absolute space, which comprised that of absolute rest, made him feel uncomfortable; he realized that there seemed to be nothing in experience corresponding to this last concept. He was also not quite comfortable about the introduction of forces operating at a distance. But the tremendous practical success of his doctrines may well have prevented him and the physicists of the eighteenth and nineteenth centuries from recognizing the fictitious character of the foundations of his system.

The natural philosophers of those days were, on the contrary, most of them possessed with the idea that the fundamental concepts and postulates of physics were not in the logical sense free inventions of the human mind but could be deduced from experience by "abstraction"—that is to say, by logical means. A clear recognition of the erroneousness of this notion really only came with the general theory of relativity, which showed that one could take account of a wider range of empirical facts, and that, too, in a more satisfactory and complete manner, on a foundation quite different from the Newtonian. But quite apart from the question of the superiority of one or the other, the fictitious character of fundamental principles is perfectly evident from the fact that we can point to two essentially different principles, both of which correspond with experience to a large extent; this proves at the same time that every attempt at a logical deduction of the basic concepts and postulates of mechanics from elementary experiences is doomed to failure.

If, then, it is true that the axiomatic basis of theoretical physics cannot be extracted from experience but must be freely invented, can we ever hope to find the right way? Nay, more, has this right way any existence outside our illusions? Can we hope to be guided safely by experience at all when there exist theories (such as classical mechanics) which to a large extent do justice to experience, without getting to the root of the matter? I answer without hesitation that there is, in my opinion, a right way, and that we are capable of finding it. Our experience hitherto justifies us in believing that nature is the realization of the simplest conceivable mathematical ideas. I am convinced that we can discover by means of purely mathematical constructions the concepts and the laws connecting them with each other, which furnish the key to the understanding of natural phenomena. Experience may suggest the appropriate mathematical concepts, but they most certainly cannot be deduced from it. Experience remains, of course, the sole criterion of the physical utility of a mathematical construction. But the creative principle resides in mathematics. In a certain sense, therefore, I hold it true that pure thought can grasp reality, as the ancients dreamed.

In order to justify this confidence, I am compelled to make use of a mathematical concept. The physical world is represented as a four-dimensional continuum. If I assume a Riemannian metric in it and ask what are the simplest laws which such a metric can satisfy, I arrive at the relativistic theory of gravitation in empty space. If in that space I assume a vector-field or an anti-symmetrical tensor-field which can be derived from it, and ask what are the simplest laws which such a field can satisfy, I arrive at Maxwell's equations for empty space.

At this point we still lack a theory for those parts of space in which electrical charge density does not disappear. De Broglie conjectured the existence of a wave field, which served to explain certain quantum properties of matter. Dirac found in the spinors field-magnitudes of a new sort, whose simplest equations enable one to a large extent to deduce the properties of the electron. Subsequently I discovered, in conjunction with my colleague, Dr. Walter Mayer, that these spinors form a special case of a new sort of field, mathematically connected with the four-dimensional system, which we called "semivectors." The simplest equations which such semivectors can satisfy furnish a key to the understanding of the existence of two sorts of elementary particles, of different ponderable mass and equal but opposite electrical charge. These semivectors are, after ordinary vectors, the simplest mathematical fields that are possible in a metrical continuum of four dimensions, and it looks as if they described, in a natural way, certain essential properties of electrical particles.

The important point for us to observe is that all these constructions and the laws connecting them can be arrived at by the principle of looking for the mathematically simplest concepts and the link between them. In the limited number of the mathematically existent simple field types, and the simple equations possible between them, lies the theorist's hope of grasping the real in all its depth.

Meanwhile the great stumbling-block for a field-theory of this kind lies in the conception of the atomic structure of matter and energy. For the theory is fundamentally non-atomic in so far as it operates exclusively with continuous functions of space, in contrast to classical mechanics, whose most important element, the material point, in itself does justice to the atomic structure of matter.

The modern quantum theory in the form associated with the names of de Broglie, Schrödinger, and Dirac, which operates with continuous functions, has overcome these difficulties by a bold piece of interpretation which was first given a clear form by Max Born. According to this, the spatial functions which appear in the equations make no claim to be a mathematical model of the atomic structure. Those functions are only supposed to determine the mathematical probabilities to find such structures, if measurements are taken, at a particular spot or in a certain state of motion. This notion is logically unobjectionable and has important successes to its credit. Unfortunately, however, it compels one to use a continuum the number of whose dimensions is not that ascribed to space by physics hitherto (four) but rises indefinitely with the number of the particles constituting the system under consideration. I cannot but confess that I attach only a transitory importance to this interpretation. I still believe in the possibility of a model of reality—that is to say, of a theory which represents things themselves and not merely the probability of their occurrence.

On the other hand, it seems to me certain that we must give up the idea of a complete localization of the particles in a theoretical model. This seems to me to be the permanent upshot of Heisenberg's principle of uncertainty. But an atomic theory in the true sense of the word (not merely on the basis of an interpretation) without localization of particles in a mathematical model is perfectly thinkable. For instance, to account for the atomic character of electricity, the field equations need only lead to the following conclusions: A region of three-dimensional space at whose boundary electrical density vanishes everywhere always contains a total electrical charge whose size is represented by a whole number. In a continuum-theory atomic characteristics would be satisfactorily expressed by integral laws without localization of the entities which constitute the atomic structure.

Not until the atomic structure has been successfully represented in such a manner would I consider the quantum-riddle solved.

THE PROBLEM OF SPACE, ETHER, AND THE FIELD IN PHYSICS

Mein Weltbild, *Amsterdam: Querido Verlag, 1934.*

Scientific thought is a development of pre-scientific thought. As the concept of space was already fundamental in the latter, we must begin with the concept of space in pre-scientific thought. There are two ways of regarding concepts, both of which are indispensable to understanding. The first is that of logical analysis. It answers the question, How do concepts and judgments depend on each other? In answering it we are on comparatively safe ground. It is the certainty by which we are so much impressed in mathematics. But this certainty is purchased at the price of emptiness of content. Concepts can only acquire content when they are connected, however indirectly, with sensible experience. But no logical investigation can reveal this connection; it can only be experienced. And yet it is this connection that determines the cognitive value of systems of concepts.

Take an example. Suppose an archaeologist belonging to a later culture finds a textbook of Euclidean geometry without diagrams. He will discover how the words "point," "straight-line," "plane" are used in the propositions. He will also recognize how the latter are deduced from each other. He will even be able to frame new propositions according to the rules he recognized. But the framing of these propositions will remain an empty play with words for him as long as "point," "straight-line," "plane," etc., convey nothing to him. Only when they do convey something will geometry possess any real content for him. The same will be true of analytical mechanics, and indeed of any exposition of a logically deductive science.

What does it mean that "straight-line," "point," "intersection," etc., convey something? It means that one can point to the sensible experiences to which those words refer. This extra-logical problem is the problem of the nature of geometry, which the archaeologist will only be able to solve intuitively by examining his experience for anything he can discover which corresponds to those primary terms of the theory and the axioms laid down for them. Only in this sense can the question of the nature of a conceptually presented entity be reasonably raised.

With our pre-scientific concepts we are very much in the position of our archaeologist in regard to the ontological problem. We have, so to speak, forgotten what features in the world of experience caused us to frame those concepts, and we have great difficulty in calling to mind the world of experience without the spectacles of the old-established conceptual interpretation. There is the further difficulty that our language is compelled to work with words which are inseparably connected with those primitive concepts. These are the obstacles which confront us when we try to describe the essential nature of the pre-scientific concept of space.

One remark about concepts in general, before we turn to the problem of space: concepts have reference to sensible experience, but they are never, in a logical sense, deducible from them. For this reason I have never been able to understand the quest of the *a priori* in the Kantian sense. In any ontological question, our concern can only be to seek out those characteristics in the complex of sense experiences to which the concepts refer.

Now as regards the concept of space: this seems to presuppose the concept of the solid body. The nature of the complexes and sense-impressions which are probably responsible for that concept has often been described. The correspondence between certain visual and tactile impressions, the fact that they can be continuously followed through time, and that the impressions can be repeated at any moment (touch, sight), are some of those characteristics. Once the concept of the solid body is formed in connection with the experiences just mentioned—which concept by no means presupposes that of space or spatial relation—the desire to get an intellectual grasp of the relations of such solid bodies is bound to give rise to concepts which correspond to their spatial relations. Two solid bodies may touch one another or be distant from one another. In the latter case, a third body can be inserted between them without altering them in any way; in the former, not. These spatial relations are obviously real in the same sense as the bodies themselves. If two bodies are equivalent with respect to filling out *one* such interval, they will also prove equivalent for other intervals. The interval is thus shown to be independent of the selection of any special body to fill it; the same is universally true of spatial relations. It is evident that this independence, which is a principal condition of the usefulness of framing purely geometrical concepts, is not necessary *a priori*. In my opinion, this concept of the interval, detached as it is from the selection of any special body to occupy it, is the starting point of the whole concept of space.

Considered, then, from the point of view of sense experience, the development of the concept of space seems, after these brief indications, to conform to the following schema—solid body; spatial relations of solid bodies; interval; space. Looked at in this way, space appears as something real in the same sense as solid bodies.

It is clear that the concept of space as a real thing already existed in the extra-scientific conceptual world. Euclid's mathematics, however, knew nothing of this concept as such; it confined itself to the concepts of the object, and the spatial relations between objects. The point, the plane, the straight line, the segment are solid objects idealized. All spatial relations are reduced to those of contact (the intersection of straight lines and planes, points lying on straight lines, etc.). Space as a continuum does not figure in the conceptual system at all. This concept was first introduced by Descartes, when he described the point-in-space by its coordinates. Here for the first time geometrical figures appear, in a way, as parts of infinite space, which is conceived as a three-dimensional continuum.

The great superiority of the Cartesian treatment of space is by no means confined to the fact that it applies analysis to the purposes of geometry. The main point seems rather to be this: the Greeks favor in their geometrical descriptions particular objects (the straight line, the plane); other objects (e.g., the ellipse) are only accessible to this description by a construction or definition with the help of the point, the straight line, and the plane. In the Cartesian treatment, on the other hand, all surfaces, for example, appear, in principle, on equal footing, without any arbitrary preference for linear structures in building up geometry.

In so far as geometry is conceived as the science of laws governing the mutual spatial relations of practically rigid bodies, it is to be regarded as the oldest branch of physics. This science was able, as I have already observed, to get along without the concept of space as such, the ideal corporeal forms—point, straight line, plane, segment—being sufficient for its needs. On the other hand, space as a whole, as conceived by Descartes, was absolutely necessary to Newtonian physics. For dynamics cannot manage with the concepts of the mass point and the (temporally variable) distance between mass points alone. In Newton's equations of motion, the concept of acceleration plays a fundamental part, which cannot be defined by the temporally variable intervals between points alone. Newton's acceleration is only conceivable or definable in relation to space as a whole. Thus to the geometrical reality of the concept of space a new inertia-determining function of space was added. When Newton described space as absolute, he no doubt meant this real significance of space, which made it necessary for him to attribute to it a quite definite state of motion, which yet did not appear to be fully determined by the phenomena of mechanics. This space was conceived as absolute in another sense also; its inertia-determining effect was conceived as autonomous, i.e., not to be influenced by any physical circumstance whatever; it affected masses, but nothing affected it.

And yet in the minds of physicists space remained until the most recent time simply the passive container of all events, without taking any part in physical occurrences. Thought only began to take a new turn with the wave-theory of light and the theory of the electromagnetic field of Faraday and Maxwell. It became clear that there existed in free space states which propagated themselves in waves, as well as localized fields which were able to exert forces on electrical masses or magnetic poles brought to the spot. Since it would have seemed utterly absurd to the physicists of the nineteenth century to attribute physical functions or states to space itself, they invented a medium pervading the whole of space, on the model of ponderable matter —the ether, which was supposed to act as a vehicle for electromagnetic phenomena, and hence for those of light also. The states of this medium, imagined as constituting the electromagnetic fields, were at first thought of mechanically, on the model of the elastic deformations of solid bodies. But this mechanical theory of the ether was never quite successful so that gradually a more detailed interpretation of the nature of etheric fields was given up. The ether thus became a kind of matter whose only function was to act as a substratum for electrical fields which were by their very nature not further analyzable. The picture was, then, as follows: space is filled by the ether, in which the material corpuscles or atoms of ponderable matter swim around; the atomic structure of the latter had been securely established by the turn of the century.

Since the interaction of bodies was supposed to be accomplished through fields, there had also to be a gravitational field in the ether, whose field-law had, however, assumed no clear form at that time. The ether was only supposed to be the seat of all forces acting across space. Since it had been realized that electrical masses in motion produce a magnetic field, whose energy provided a model for inertia, inertia also appeared as a field-action localized in the ether.

The mechanical properties of the ether were at first a mystery. Then came H. A. Lorentz's great discovery. All the phenomena of electromagnetism then known could be explained on the basis of two assumptions: that the ether is firmly fixed in space— that is to say, unable to move at all, and that electricity is firmly lodged in the mobile elementary particles. Today his discovery may be expressed as follows: physical space and the ether are only different terms for the same thing; fields are physical states of space. For if no particular state of motion can be ascribed to the ether, there does not seem to be any ground for introducing it as an entity of a special sort alongside of space. But the physicists were still far removed from such a way of thinking; space was still, for them, a rigid, homogeneous something, incapable of changing or assuming various states. Only the genius of Riemann, solitary and uncomprehended, had already won its way by the middle of the last century to a new conception of space, in which space was deprived of its rigidity, and the possibility of its partaking in physical events was recognized. This intellectual achievement commands our admiration all the more for having preceded Faraday's and Maxwell's field theory of electricity. Then came the special theory of relativity with its recognition of the physical equivalence of all inertial systems. The inseparability of time and space emerged in connection with electrodynamics, or the, law of the propagation of light. Hitherto it had been silently assumed that the four-dimensional continuum of events could be split up into time and space in an objective manner—i.e., that an absolute significance attached to the "now" in the world of events. With the discovery of the relativity of simultaneity, space and time were merged in a single continuum in a way similar to that in which the three dimensions of space had previously been merged into a single continuum. Physical space was thus extended to a four-dimensional space which also included the dimension of time. The four-dimensional space of the special theory of relativity is just as rigid and absolute as Newton's space.

The theory of relativity is a fine example of the fundamental character of the modern development of theoretical science. The initial hypotheses become steadily more abstract and remote from experience. On the other hand, it gets nearer to the grand aim of all science, which is to cover the greatest possible number of empirical facts by logical deduction from the smallest possible number of hypotheses or axioms. Meanwhile, the train of thought leading from the axioms to the empirical facts or verifiable consequences gets steadily longer and more subtle. The theoretical scientist is compelled in an increasing degree to be guided by purely mathematical, formal considerations in his search for a theory, because the physical experience of the experimenter cannot lead him up to the regions of highest abstraction. The predominantly inductive methods appropriate to the youth of science are giving place to tentative deduction. Such a theoretical structure needs to be very thoroughly elaborated before it can lead to conclusions which can be compared with experience. Here, too, the observed fact is undoubtedly the supreme arbiter; but it cannot pronounce sentence until the wide chasm separating the axioms from their verifiable consequences has been bridged by much intense, hard thinking. The theorist has to set about this Herculean task fully aware that his efforts may only be destined to prepare the death blow to his theory. The theorist who undertakes such a labor should not be carped at as "fanciful"; on the contrary, he should be granted the right to give free reign to his fancy, for there is no other way to the goal. His is no idle daydreaming, but a search for the logically simplest possibilities and their consequences. This plea was needed in order to make the listener or reader more inclined to follow the ensuing train of ideas with attention; it is the line of thought which has led from the special to the general theory of relativity and thence to its latest offshoot, the unified field theory. In this exposition the use of mathematical symbols cannot be completely avoided.

We start with the special theory of relativity. This theory is still based directly on an empirical law, that of the constancy of the velocity of light. Let P be a point in empty space, P' an infinitely close point at a distance $d\sigma$. Let a flash of light be emitted from P at a time t and reach P' at a time $t + dt$. Then

$$d\sigma^2 = c^2 dt^2$$

If dx_1, dx_2, dx_3 are the orthogonal projections of $d\sigma$, and the imaginary time coordinate $\sqrt{-1}\,ct = x_4$ is introduced, then the above-mentioned law of the constancy of the velocity of light propagation takes the form

$$ds^2 = dx_1^2 + dx_2^2 + dx_3^2 + dx_4^2 = 0$$

Since this formula expresses a real situation, we may attribute a real meaning to the quantity ds, even if the neighboring points of the four-dimensional continuum are so chosen that the corresponding ds does not vanish. This may be expressed by saying that the four-dimensional space (with an imaginary time-coordinate) of the special theory of relativity possesses a Euclidean metric.

The fact that such a metric is called Euclidean is connected with the following. The postulation of such a metric in a three-dimensional continuum is fully equivalent to the postulation of the axioms of Euclidean geometry. The defining equation of the metric is then nothing but the Pythagorean theorem applied to the differentials of the coordinates.

In the special theory of relativity those coordinate changes (by transformation) are permitted for which also in the new coordinate system the quantity ds^2 (fundamental invariant) equals the sum of the squares of the coordinate differentials. Such transformations are called Lorentz transformations.

The heuristic method of the special theory of relativity is characterized by the following principle: only those equations are admissible as an expression of natural laws which do not change their form when the coordinates are changed by means of a Lorentz transformation (covariance of equations with respect to Lorentz transformations).

This method led to the discovery of the necessary connection between momentum and energy, between electric and magnetic field strength, electrostatic and electrodynamic forces, inert mass and energy; and the number of independent concepts and fundamental equations in physics was thereby reduced.

This method pointed beyond itself. Is it true that the equations which express natural laws are covariant with respect to Lorentz transformations only and not with respect to other transformations? Well, formulated in that way the question really has no meaning, since every system of equations can be expressed in general coordinates. We must ask: Are not the laws of nature so constituted that they are not materially simplified through the choice of any one *particular* set of coordinates?

We will only mention in passing that our empirical law of the equality of inert and gravitational masses prompts us to answer this question in the affirmative. If we elevate the equivalence of all coordinate systems for the formulation of natural laws into a principle, we arrive at the general theory of relativity, provided we retain the law of the constancy of the velocity of light or, in other words, the hypothesis of the objective significance of the Euclidean metric at least for infinitely small portions of four-dimensional space.

This means that for finite regions of space the (physically meaningful) existence of a general Riemannian metric is postulated according to the formula

$$ds^2 = \sum_{\mu\nu} g_{\mu\nu} \, dx_\mu \, dx_\nu,$$

where the summation is to be extended to all index combinations from 1,1 to 4,4.

The structure of such a space differs quite basically in *one* respect from that of a Euclidean space. The coefficients $g_{\mu\nu}$ are for the time being any functions whatever of the coordinates x_1 to x_4, and the structure of the space is not really determined until these functions $g_{\mu\nu}$ are really known. One can also say: the structure of such a space is as such completely undetermined. It is only determined more closely by specifying laws which the metrical field of the $g_{\mu\nu}$ satisfy. On physical grounds it was assumed that the metrical field was at the same time the gravitational field.

Since the gravitational field is determined by the configuration of masses and changes with it, the geometric structure of this space is also dependent on physical factors. Thus, according to this theory space is—exactly as Riemann guessed—no longer absolute; its structure depends on physical influences. (Physical) geometry is no longer an isolated self-contained science like the geometry of Euclid.

The problem of gravitation was thus reduced to a mathematical problem: it was required to find the simplest fundamental equations which are covariant with respect to arbitrary coordinate transformation. This was a well-defined problem that could at least be solved.

I will not speak here of the experimental confirmation of this theory, but explain at once why the theory could not rest permanently satisfied with this success. Gravitation had indeed been deduced from the structure of space, but besides the gravitational field there is also the electromagnetic field. This had, to begin with, to be introduced into the theory as an entity independent of gravitation. Terms which took account of the existence of the electromagnetic field had to be added to the fundamental field equations. But the idea that there exist two structures of space independent of each other, the metric-gravitational and the electromagnetic, was intolerable to the theoretical spirit. We are prompted to the belief that both sorts of field must correspond to a unified structure of space.

NOTES ON THE ORIGIN OF THE GENERAL THEORY OF RELATIVITY

Mein Weltbild, *Amsterdam: Querido Verlag, 1934.*

I gladly accede to the request that I should say something about the history of my own scientific work. Not that I have an exaggerated notion of the importance of my own efforts, but to write the history of other men's work demands a degree of absorption in other people's ideas which is much more in the line of the trained historian; to throw light on one's own earlier thinking appears incomparably easier. Here one has an immense advantage over everybody else, and one ought not to leave the opportunity unused out of modesty.

When by the special theory of relativity I had arrived at the equivalence of all so-called inertial systems for the formulation of natural laws (1905), the question whether there was not a further equivalence of coordinate systems followed naturally, to say the least of it. To put it in another way, if only a relative meaning can be attached to the concept of velocity, ought we nevertheless to persevere in treating acceleration as an absolute concept?

From the purely kinematic point of view there was no doubt about the relativity of all motions whatever; but physically speaking, the inertial system seemed to occupy a privileged position, which made the use of coordinate systems moving in other ways appear artificial.

I was of course acquainted with Mach's view, according to which it appeared conceivable that what inertial resistance counteracts is not acceleration as such but acceleration with respect to the masses of the other bodies existing in the world. There was something fascinating about this idea to me, but it provided no workable basis for a new theory.

I first came a step nearer to the solution of the problem when I attempted to deal with the law of gravity within the framework of the special theory of relativity. Like most writers at the time, I tried to frame a *field-law* for gravitation, since it was no longer possible, at least in any natural way, to introduce direct action at a distance owing to the abolition of the notion of absolute simultaneity.

The simplest thing was, of course, to retain the Laplacian scalar potential of gravity, and to complete the equation of Poisson in an obvious way by a term differentiated with respect to time in such a way that the special theory of relativity was satisfied. The law of motion of the mass point in a gravitational field had also to be adapted to the special theory of relativity. The path was not so unmistakably marked out here, since the inert mass of a body might depend on the gravitational potential. In fact, this was to be expected on account of the principle of the inertia of energy.

These investigations, however, led to a result which raised my strong suspicions. According to classical mechanics, the vertical acceleration of a body in the vertical gravitational field is independent of the horizontal component of its velocity. Hence in such a gravitational field the vertical acceleration of a mechanical system or of its center of gravity works out independently of its internal kinetic energy. But in the theory I advanced, the acceleration of a falling body was not independent of its horizontal velocity or the internal energy of a system.

This did not fit in with the old experimental fact that all bodies have the same acceleration in a gravitational field. This law, which may also be formulated as the law of the equality of inertial and gravitational mass, was now brought home to me in all its significance. I was in the highest degree amazed at its existence and guessed that in it must lie the key to a deeper understanding of inertia and gravitation. I had no serious doubts about its strict validity even without knowing the results of the admirable experiments of Eötvös, which—if my memory is right—I only came to know later. I now abandoned as inadequate the attempt to treat the problem of gravitation, in the manner outlined above, within the framework of the special theory of relativity. It clearly failed to do justice to the most fundamental property of gravitation. The principle of the equality of inertial and gravitational mass could now be formulated quite clearly as follows: In a homogeneous gravitational field all motions take place in the same way as in the absence of a gravitational field in relation to a uniformly accelerated coordinate system. If this principle held good for any events whatever (the "principle of equivalence"), this was an indication that the principle of relativity needed to be extended to coordinate systems in non-uniform motion with respect to each other, if we were to reach a natural theory of the gravitational fields. Such reflections kept me busy from 1908 to 1911, and I attempted to draw special conclusions from them, of which I do not propose to speak here. For the moment the one important thing was the discovery that a reasonable theory of gravitation could only be hoped for from an extension of the principle of relativity.

What was needed, therefore, was to frame a theory whose equations kept their form in the case of non-linear transformations of the coordinates. Whether this was to apply to arbitrary (continuous) transformations of coordinates or only to certain ones, I could not for the moment say.

I soon saw that the inclusion of non-linear transformations, as the principle of equivalence demanded, was inevitably fatal to the simple physical interpretation of the coordinates—i.e., that it could no longer be required that coordinate differences should signify direct results of measurement with ideal scales or clocks. I was much bothered by this piece of knowledge, for it took me a long time to see what coordinates at all meant in physics. I did not find the way out of this dilemma until 1912, and then it came to me as a result of the following consideration:

A new formulation of the law of inertia had to be found which in case of the absence of a "real gravitational field" passed over into Galileo's formulation for the principle of inertia if an inertial system was used as coordinate system. Galileo's formulation amounts to this: A material point, which is acted on by no force, will be represented in four-dimensional space by a straight line, that is to say, by a shortest line, or more correctly, an extremal line. This concept presupposes that of the length of a line element, that is to say, a metric. In the special theory of relativity, as Minkowski had shown, this metric was a quasi-Euclidean one, i.e., the square of the "length" ds of a line element was a certain quadratic function of the differentials of the coordinates.

If other coordinates are introduced by means of a non-linear transformation, ds^2 remains a homogeneous function of the differentials of the coordinates, but the coefficients of this function $(g_{\mu\nu})$ cease to be constant and become certain functions of the coordinates. In mathematical terms this means that physical (four-dimensional) space has a Riemannian metric. The timelike extremal lines of this metric furnish the law of motion of a material point which is acted on by no force apart from the forces of gravity. The coefficients $(g_{\mu\nu})$ of this metric at the same time describe the gravitational field with reference to the coordinate system selected. A natural formulation of the principle of equivalence had thus been found, the extension of which to any gravitational field whatever formed a perfectly natural hypothesis.

The solution of the above-mentioned dilemma was therefore as follows: A physical significance attaches not to the differentials of the coordinates but only to the Riemannian metric corresponding to them. A workable basis had now been found for the general theory of relativity. Two further problems remained to be solved, however.

1. If a field-law is expressed in terms of the special theory of relativity, how can it be transferred to the case of a Riemannian metric?

2. What are the differential laws which determine the Riemannian metric (i.e., $g_{\mu\nu}$) itself?

I worked on these problems from 1912 to 1914 together with my friend Grossmann. We found that the mathematical methods for solving problem 1 lay ready in our hands in the absolute differential calculus of Ricci and Levi-Civita.

As for problem 2, its solution obviously required the construction (from the $g_{\mu\nu}$) of the differential invariants of the second order. We soon saw that these had already been established by Riemann (the tensor of curvature). We had already considered the right field-equation for gravitation two years before the publication of the general theory of relativity, but we were unable to see how they could be used in physics. On the contrary, I felt sure that they could not do justice to experience. Moreover I believed that I could show on general considerations that a law of gravitation invariant with respect to arbitrary transformations of coordinates was inconsistent with the principle of causality. These were errors of thought which cost me two years of excessively hard work, until I finally recognized them as such at the end of 1915, and after having ruefully returned to the Riemannian curvature, succeeded in linking the theory with the facts of astronomical experience.

In the light of knowledge attained, the happy achievement seems almost a matter of course, and any intelligent student can grasp it without too much trouble. But the years of anxious searching in the dark, with their intense longing, their alternations of confidence and exhaustion and the final emergence into the light—only those who have experienced it can understand that.

PHYSICS AND REALITY

From The Journal of the Franklin Institute, *Vol. 221, No. 3. March, 1936.*

I. GENERAL CONSIDERATION CONCERNING THE METHOD OF SCIENCE

It has often been said, and certainly not without justification, that the man of science is a poor philosopher. Why, then, should it not be the right thing for the physicist to let the philosopher do the philosophizing? Such might indeed be the right thing at a time when the physicist believes he has at his disposal a rigid system of fundamental concepts and fundamental laws which are so well established that waves of doubt cannot reach them; but, it cannot be right at a time when the very foundations of physics itself have become problematic as they are now. At a time like the present, when experience forces us to seek a newer and more solid foundation, the physicist cannot simply surrender to the philosopher the critical contemplation of the theoretical foundations; for, he himself knows best, and feels more surely where the shoe pinches. In looking for a new foundation, he must try to make clear in his own mind just how far the concepts which he uses are justified, and are necessities.

The whole of science is nothing more than a refinement of everyday thinking. It is for this reason that the critical thinking of the physicist cannot possibly be restricted to the examination of the concepts of his own specific field. He cannot proceed without considering critically a much more difficult problem, the problem of analyzing the nature of everyday thinking.

Our ·psychological experience contains, in colorful succession, sense experiences, memory pictures of them, images, and feelings. In contrast to psychology, physics treats directly only of sense experiences and of the "understanding" of their connection. But even the concept of the "real external world" of everyday thinking rests exclusively on sense impressions.

Now we must first remark that the differentiation between sense impressions and images is not possible; or, at least it is not possible with absolute certainty. With the discussion of this problem, which affects also the notion of reality, we will not concern ourselves but we shall take the existence of sense experiences as given, that is to say, as psychic experiences of a special kind.

I believe that the first step in the setting of a "real external world" is the formation of the concept of bodily objects and of bodily objects of various kinds. Out of the multitude of our sense experiences we take, mentally and arbitrarily, certain repeatedly occurring complexes of sense impressions (partly in conjunction with sense impressions which are interpreted as signs for sense experiences of others), and we correlate to them a concept—the concept of the bodily object. Considered logically this concept is not identical with the totality of sense impressions referred to; but it is a free creation of the human (or animal) mind. On the other hand, this concept owes its meaning and its justification exclusively to the totality of the sense impressions which we associate with it.

The second step is to be found in the fact that, in our thinking (which determines our expectation), we attribute to this concept of the bodily object a significance, which is to a high degree independent of the sense impressions which originally give rise to it. This is what we mean when we attribute to the bodily object "a real existence." The justification of such a setting rests exclusively on the fact that, by means of such concepts and mental relations between them, we are able to orient ourselves in the labyrinth of sense impressions. These notions and relations, although free mental creations, appear to us as stronger and more unalterable than the individual sense experience itself, the character of which as anything other than the result of an illusion or hallucination is never completely guaranteed. On the other hand, these concepts and relations, and indeed the postulation of real objects and, generally speaking, of the existence of "the real world," have justification only in so far as they are connected with sense impressions between which they form a mental connection.

The very fact that the totality of our sense experiences is such that by means of thinking (operations with concepts, and the creation and use of definite functional relations between them, and the coordination of sense experiences to these concepts) it can be put in order, this fact is one which leaves us in awe, but which we shall never understand. One may say "the eternal mystery of the world is its comprehensibility." It is one of the great realizations of Immanuel Kant that the postulation of a real external world would be senseless without this comprehensibility.

In speaking here of "comprehensibility," the expression is used in its most modest sense. It implies: the production of some sort of order among sense impressions, this order being produced by the creation of general concepts, relations between these concepts, and by definite relations of some kind between the concepts and sense experience. It is in this sense that the world of our sense experiences is comprehensible. The fact that it is comprehensible is a miracle.

In my opinion, nothing can be said *a priori* concerning the manner in which the concepts are to be formed and connected, and how we are to coordinate them to sense experiences. In guiding us in the creation of such an order of sense experiences, success alone is the determining factor. All that is necessary is to fix a set of rules, since without such rules the acquisition of knowledge in the desired sense would be impossible. One may compare these rules with the rules of a game in which, while the rules themselves are arbitrary, it is their rigidity alone which makes the game possible. However, the fixation will never be final. It will have validity only for a special field of application (i.e., there are no final categories in the sense of Kant).

The connection of the elementary concepts of everyday thinking with complexes of sense experiences can only be comprehended intuitively and it is unadaptable to scientifically logical fixation. The totality of these connections—none of which is expressible in conceptual terms—is the only thing which differentiates the great building which is science from a logical but empty scheme of concepts. By means of these connections, the purely conceptual propositions of science become general statements about complexes of sense experiences.

We shall call "primary concepts" such concepts as are directly and intuitively connected with typical complexes of sense experiences. All other notions are—from the physical point of view—possessed of meaning only in so far as they are connected, by propositions, with the primary notions. These propositions are partially definitions of the concepts (and of the statements derived logically from them) and partially propositions not derivable from the definitions, which express at least indirect relations between the "primary concepts," and in this way between sense experiences. Propositions of the latter kind are "statements about reality" or laws of nature, i.e.. propositions which have to show their validity when applied to sense experiences covered by primary concepts. The question as to which of the propositions shall be considered as definitions and which as natural laws will depend largely upon the chosen representation. It really becomes absolutely necessary to make this differentiation only when one examines the degree to which the whole system of concepts considered is not empty from the physical point of view.

STRATIFICATION OF THE SCIENTIFIC SYSTEM

The aim of science is, on the one hand, a comprehension, as *complete* as possible, of the connection between the sense experiences in their totality, and, on the other hand, the accomplishment of this aim *by the use of a minimum of primary concepts and relations*. (Seeking, as far as possible, logical unity in the world picture, i.e., paucity in logical elements.)

Science uses the totality of the primary concepts, i.e., concepts directly connected with sense experiences, and propositions connecting them. In its first stage of development, science does not contain anything else. Our everyday thinking is satisfied on the whole with this level. Such a state of affairs cannot, however, satisfy a spirit which is really scientifically minded; because the totality of concepts and relations obtained in this manner is utterly lacking in logical unity. In order to supplement this deficiency, one invents a system poorer in concepts and relations, a system retaining the primary concepts and relations of the "first layer" as logically derived concepts and relations. This new "secondary system" pays for its higher logical unity by having elementary concepts (concepts of the second layer), which are no longer directly connected with complexes of sense experiences. Further striving for logical unity brings us to a tertiary system, still poorer in concepts and relations, for the deduction of the concepts and relations of the secondary (and so indirectly of the primary) layer. Thus the story goes on until we have arrived at a system of the greatest conceivable unity, and of the greatest poverty of concepts of the logical foundations, which is still compatible with the observations made by our senses. We do not know whether or not this ambition will ever result in a definitive system. If one is asked for his opinion, he is inclined to answer no. While wrestling with the problems, however, one will never give up the hope that this greatest of all aims can really be attained to a very high degree.

An adherent to the theory of abstraction or induction might call our layers "degrees of abstraction"; but I do not consider it justifiable to veil the logical independence of the concept from the sense experiences. The relation is not analogous to that of soup to beef but rather of check number to overcoat.

The layers are furthermore not clearly separated. It is not even absolutely clear which concepts belong to the primary layer. As a matter of fact, we are dealing with freely formed concepts, which, with a certainty sufficient for practical use, are intuitively connected with complexes of sense experiences in such a manner that, in any given case of experience, there is no uncertainty as to the validity of an assertion. The essential thing is the aim to represent the multitude of concepts and propositions, close to experience, as propositions, logically deduced from a basis, as narrow as possible, of fundamental concepts and fundamental relations which themselves can be chosen freely (axioms). The liberty of choice, however, is of a special kind; it is not in any way similar to the liberty of a writer of fiction. Rather, it is similar to that of a man engaged in solving a well-designed word puzzle. He may, it is true, propose any word as the solution; but, there is only *one* word which really solves the puzzle in all its parts. It is a matter of faith that nature —as she is perceptible to our five senses—takes the character of such a well-formulated puzzle. The successes reaped up to now by science do, it is true, give a certain encouragement for this faith.

The multitude of layers discussed above corresponds to the several stages of progress which have resulted from the struggle for unity in the course of development. As regards the final aim, intermediary layers are only of temporary nature. They must eventually disappear as irrelevant. We have to deal, however, with the science of today, in which these strata represent problematic partial successes which support one another but which also threaten one another, because today's system of concepts contains deep-seated incongruities which we shall meet later on.

It will be the aim of the following lines to demonstrate what paths the constructive human mind has entered, in order to arrive at a basis of physics which is logically as uniform as possible.

II. Mechanics and the Attempts to Base All Physics upon It

An important property of our sense experiences, and, more generally, of all of our experiences, is their temporal order. This kind of order leads to the mental conception of a subjective time, an ordering scheme for our experience. The subjective time leads then via the concept of the bodily object and of space to the concept of objective time, as we shall see later on.

Ahead of the notion of objective time there is, however, the concept of space; and ahead of the latter we find the concept of the bodily object. The latter is directly connected with complexes of sense experiences. It has been pointed out that one property which is characteristic of the notion "bodily object" is the property which provides that we coordinate to it an existence, independent of (subjective) time, and independent of the fact that it is perceived by our senses. We do this in spite of the fact that we perceive temporal alterations in it. Poincaré has justly emphasized the fact that we distinguish two kinds of alterations of the bodily object, "changes of state" and "changes of position." The latter, he remarked, are alterations which we can reverse by voluntary motions of our bodies.

That there are bodily objects to which we have to ascribe, within a certain sphere of perception, no alteration of state, but only alterations of position, is a fact of fundamental importance for the formation of the concept of space (in a certain degree even for the justification of the notion of the bodily object itself). Let us call such an object "practically rigid."

If, as the object of our perception, we consider simultaneously (i.e., as a single unit) two practically rigid bodies, then there exist for this ensemble such alterations as can *not* possibly be considered as changes of position of the whole, notwithstanding the fact that this is the case for each one of the two constituents. This leads to the notion of "change of relative position" of the two objects; and, in this way, also to the notion of "relative position" of the two objects. It is found moreover that among the relative positions, there is one of a specific kind which we designate as "contact." * Permanent contact of two bodies in three or more "points" means that they are united to a quasi-rigid compound body. It is permissible to say that the second body forms then a (quasi-rigid) continuation of the first body and may, in its turn, be continued quasi-rigidly. The possibility of the quasi-rigid continuation of a body is unlimited. The totality of all conceivable quasi-rigid continuations of a body B_0 is the infinite "space" determined by it.

In my opinion, the fact that every bodily object situated in any arbitrary manner can be put into contact with the quasi-rigid continuation of some given body B_0 (body of reference), this fact is the empirical basis of our conception of space. In pre-scientific thinking, the solid earth's crust plays the role of B_0 and its continuation. The very name geometry indicates

---

* It is in the nature of things that we are able to talk about these objects only by means of concepts of our own creation, concepts which themselves are not subject to definition. It is essential, however, that we make use only of such concepts concerning whose coordination to our experience we feel no doubt.

that the concept of space is psychologically connected with the earth as an ever present body of reference.

The bold notion of "space" which preceded all scientific geometry transformed our mental concept of the relations of positions of bodily objects into the notion of the position of these bodily objects in "space." This, of itself, represents a great formal simplification. Through this concept of space one reaches, moreover, an attitude in which any description of position is implicitly a description of contact; the statement that a point of a bodily object is located at a point P of space means that the object touches the point P of the standard body of reference B_0 (supposed appropriately continued) at the point considered.

In the geometry of the Greeks, space plays only a qualitative role, since the position of bodies in relation to space is considered as given, it is true, but is not described by means of numbers. Descartes was the first to introduce this method. In his language, the whole content of Euclidean geometry can axiomatically be founded upon the following statements: (1) Two specified points of a rigid body determine a segment. (2) We may associate triples of numbers X_1, X_2, X_3, to points of space in such a manner that for every segment $P' - P''$ under consideration, the coordinates of whose end points are X_1', X_2', X_3'; X_1'', X_2'', X_3'', the expression

$$s^2 = (X_1'' - X_1')^2 + (X_2'' - X_2')^2 + (X_3'' - X_3')^2$$

is independent of the position of the body, and of the positions of any and all other bodies.

The (positive) number s is called the length of the segment, or the distance between the two points P' and P'' of space (which are coincident with the points P' and P'' of the segment).

The formulation is chosen, intentionally, in such a way that it expresses clearly, not only the logical and axiomatic, but also the empirical content of Euclidean geometry. The purely logical (axiomatic) representation of Euclidean geometry has, it is true, the advantage of greater simplicity and clarity. It pays for this, however, by renouncing a representation of the connection between the conceptual construction and the sense experiences upon which connection, alone, the significance of geometry for physics rests. The fatal error that logical necessity, preceding all experience, was the basis of Euclidean geometry and the concept of space belonging to it, this fatal error arose from the fact that the empirical basis, on which the axiomatic construction of Euclidean geometry rests, had fallen into oblivion.

In so far as one can speak of the existence of rigid bodies in nature, Euclidean geometry is a physical science, which must be confirmed by sense experiences. It concerns the totality of laws which must hold for the relative positions of rigid bodies independently of time. As one may see, the physical notion of space also, as originally used in physics, is tied to the existence of rigid bodies.

From the physicist's point of view, the central importance of Euclidean geometry rests in the fact that its laws are independent of the specific nature of the bodies whose relative positions it discusses. Its formal simplicity is characterized by the properties of homogeneity and isotropy (and the existence of similar entities).

The concept of space is, it is true, useful, but not indispensable for geometry proper, i.e., for the formulation of rules about the relative positions of rigid bodies. By contrast, the concept of objective time, without which the formulation of the fundamentals of classical mechanics is impossible, is linked with the concept of the spatial continuum.

The introduction of objective time involves two postulates which are independent of each other.

1. The introduction of the objective local time by connecting the temporal sequence of experiences with the readings of a "clock," i.e., of a periodically recurring closed system.

2. The introduction of the notion of objective time for the events in the whole space, by which notion alone the idea of local time is extended to the idea of time in physics.

Note concerning 1. As I see it, it does not mean a "petitio principii" if one puts the concept of periodical recurrence ahead of the concept of time, while one is concerned with the clarification of the origin and of the empirical content of the concept of time. Such a conception corresponds exactly to the precedence of the concept of the rigid (or quasi-rigid) body in the interpretation of the concept of space.

Further discussion of 2. The illusion which prevailed prior to the enunciation of the theory of relativity—that, from the point of view of experience the meaning of simultaneity in relation to spatially distant events and, consequently, that the meaning of physical time is *a priori* clear—this illusion had its origin in the fact that in our everyday experience we can neglect the time of propagation of light. We are accustomed on this account to fail to differentiate between "simultaneously seen" and "simultaneously happening"; and, as a result, the difference between time and local time is blurred.

The lack of definiteness which, from the point of view of its empirical significance, adheres to the notion of time in classical mechanics was veiled by the axiomatic representation of space and time as given independently of our sense experiences. Such a use of notions—independent of the empirical basis to which they owe their existence—does not necessarily damage science. One may, however, easily be led into the error of believing that these notions, whose origin is forgotten, are logically necessary and therefore unalterable, and this error may constitute a serious danger to the progress of science.

It was fortunate for the development of mechanics and hence also for the development of physics in general, that the lack of definiteness in the concept of objective time remained hidden from the earlier philosophers as regards its empirical interpretation. Full of confidence in the real meaning of the space-time construction, they developed the foundations of mechanics which we shall characterize, schematically, as follows:

(*a*) Concept of a material point: a bodily object which—as regards its position and motion—can be described with sufficient accuracy as a point with coordinates X_1, X_2, X_3. Description of its motion (in relation to the "space" B_0) by giving X_1, X_2, X_3, as functions of the time.

(*b*) Law of inertia: the disappearance of the components of acceleration for a material point which is sufficiently far away from all other points.

(*c*) Law of motion (for the material point): Force = mass $\times$ acceleration.

(d) Laws of force (interactions between material points).

In this, (b) is merely an important special case of (c). A real theory exists only when the laws of force are given. The forces must in the first place only obey the law of equality of action and reaction in order that a system of points—permanently connected to each other by forces—may behave like *one* material point.

These fundamental laws, together with Newton's law for the gravitational force, form the basis of the mechanics of celestial bodies. In this mechanics of Newton, and in contrast to the above conceptions of space derived from rigid bodies, the space B_0 enters in a form which contains a new idea; it is not for every B_0 that validity is asserted (for a given law of force) for (b) and (c), but only for a B_0 in an appropriate state of motion (inertial system). On account of this fact, the coordinate space acquired an independent physical property which is not contained in the purely geometrical notion of space, a circumstance which gave Newton considerable food for thought (pail-experiment).*

Classical mechanics is only a general scheme; it becomes a theory only by explicit indication of the force laws (d) as was done so very successfully by Newton for celestial mechanics. From the point of view of the aim of the greatest logical simplicity of the foundations, this theoretical method is deficient in so far as the laws of force cannot be obtained by logical and formal considerations, so that their choice is *a priori* to a large extent arbitrary. Also Newton's law of gravitation is distinguished from other conceivable laws of force exclusively by its *success*.

In spite of the fact that, today, we know positively that classical mechanics fails as a foundation dominating all physics, it still occupies the center of all of our thinking in physics. The reason for this lies in the fact that, regardless of important

---

* This defect of the theory could only be eliminated by such a formulation of mechanics as would claim validity for all B_0. This is one of the steps which led to the general theory of relativity. A second defect, also eliminated only by the introduction of the general theory of relativity, lies in the fact that there is no reason given by mechanics itself for the equality of the gravitational and inertial mass of the material point.

progress reached since the time of Newton, we have not yet arrived at a new foundation of physics concerning which we may be certain that the manifold of all investigated phenomena, and of successful partial theoretical systems, could be deduced logically from it. In the following lines I shall try to describe briefly how the matter stands.

First we try to get clearly in our minds how far the system of classical mechanics has shown itself adequate to serve as a basis for the whole of physics. Since we are dealing here only with the foundations of physics and with its development, we need not concern ourselves with the purely *formal* progresses of mechanics (equations of Lagrange, canonical equations, etc.). *One* remark, however, appears indispensable. The notion "material point" is fundamental for mechanics. If now we seek to develop the mechanics of a bodily object which itself can *not* be treated as a material point—and strictly speaking every object "perceptible to our senses" is of this category—then the question arises: How shall we imagine the object to be built up out of material points, and what forces must we assume as acting between them? The formulation of this question is indispensable, if mechanics is to pretend to describe the object *completely.*

It is in line with the natural tendency of mechanics to assume these material points, and the laws of forces acting between them, as invariable, since temporal changes would lie outside of the scope of mechanical explanation. From this we can see that classical mechanics must lead us to an atomistic construction of matter. We now realize, with special clarity, how much in error are those theorists who believe that theory comes inductively from experience. Even the great Newton could not free himself from this error (*"Hypotheses non fingo"* *).

In order to save itself from becoming hopelessly lost in this line of thought (atomism), science proceeded first in the following manner. The mechanics of a system is determined if its potential energy is given as a function of its configuration. Now, if the acting forces are of such a kind as to guarantee the

---

* "I make no hypotheses."

maintenance of certain structural properties of the system's configuration, then the configuration may be described with sufficient accuracy by a relatively small number of configuration
variables q_r; the potential energy is considered only in so far as it
is dependent upon *these* variables (for instance, description of
the configuration of a practically rigid body by six variables).

A second method of application of mechanics, which avoids
the consideration of a subdivision of matter down to "real" material points, is the mechanics of so-called continuous media.
This mechanics is characterized by the fiction that the density
and the velocity of matter depend continuously upon coordinates and time, and that the part of the interactions not ex-
plicitly given can be considered as surface forces (pressure
forces) which again are continuous functions of position.
Herein we find the hydrodynamic theory, and the theory of elas-
ticity of solid bodies. These theories avoid the explicit introduction of material points by fictions which, in the light of the
foundation of classical mechanics, can only have an approximate
significance.

In addition to their great *practical* significance, these categories of science have—by developing new mathematical concepts—created those formal tools (partial differential equations)
which have been necessary for the subsequent attempts at a new
foundation of all of physics.

These two modes of application of mechanics belong to the
so-called "phenomenological" physics. It is characteristic of
this kind of physics that it makes as much use as possible of
concepts which are close to experience but, for this reason, has
to give up, to a large extent, unity in the foundations. Heat,
electricity, and light are described by separate variables of state
and material constants other than the mechanical quantities;
and to determine all of these variables in their mutual and temporal dependence was a task which, in the main, could only be
solved empirically. Many contemporaries of Maxwell saw in
such a manner of presentation the ultimate aim of physics,
which they thought could be obtained purely inductively from
experience on account of the relative closeness of the concepts

used to experience. From the point of view of theories of knowl‐
edge St. Mill and E. Mach took their stand approximately on
this ground.

In my view, the greatest achievement of Newton's mechanics
lies in the fact that its consistent application has led beyond this
phenomenological point of view, particularly in the field of heat
phenomena. This occurred in the kinetic theory of gases and
in statistical mechanics in general. The former connected the
equation of state of the ideal gases, viscosity, diffusion, and heat
conductivity of gases and radiometric phenomena of gases, and
gave the logical connection of phenomena which, from the
point of view of direct experience, had nothing whatever to do
with one another. The latter gave a mechanical interpretation
of the thermodynamic ideas and laws and led to the discovery
of the limit of applicability of the notions and laws of the classi‐
cal theory of heat. This kinetic theory, which by far surpassed
phenomenological physics as regards the logical unity of its
foundations, produced, moreover, definite values for the true
magnitudes of atoms and molecules which resulted from several
independent methods and were thus placed beyond the realm
of reasonable doubt. These decisive progresses were paid for
by the coordination of atomistic entities to the material points,
the constructively speculative character of these entities being
obvious. Nobody could hope ever to "perceive directly" an
atom. Laws concerning variables connected more directly with
experimental facts (for example: temperature, pressure, speed)
were deduced from the fundamental ideas by means of compli‐
cated calculations. In this manner physics (at least part of it),
originally more phenomenologically constructed, was reduced,
by being founded upon Newton's mechanics for atoms and
molecules, to a basis further removed from direct experiment,
but more uniform in character.

III. The Field Concept

In explaining optical and electrical phenomena, Newton's
mechanics has been far less successful than it had been in the
fields cited above. It is true that Newton tried to reduce light

to the motion of material points in his corpuscular theory of light. Later on, however, as the phenomena of polarization, diffraction, and interference of light forced upon this theory more and more unnatural modifications, Huygens' undulatory theory of light prevailed. Probably this theory owes its origin essentially to the phenomena of crystal optics and to the theory of sound, which was then already elaborated to a certain degree. It must be admitted that Huygens' theory also was based in the first instance upon classical mechanics; the all-penetrating ether had to be assumed as the carrier of the waves, but no known phenomenon suggested the way in which the ether was built up from material points. One could never get a clear picture of the internal forces governing the ether, nor of the forces acting between the ether and "ponderable" matter. The foundations of this theory remained, therefore, eternally in the dark. The true basis was a partial differential equation, the reduction of which to mechanical elements remained always problematic.

For the theoretical conception of electric and magnetic phenomena one introduced, again, masses of a special kind, and between these masses one assumed the existence of forces acting at a distance, similar to Newton's gravitational forces. This special kind of matter, however, appeared to be lacking in the fundamental property of inertia; and the forces acting between these masses and the ponderable matter remained obscure. To these difficulties there had to be added the polar character of these kinds of matter which did not fit into the scheme of classical mechanics. The basis of the theory became still more unsatisfactory when electrodynamic phenomena became known, notwithstanding the fact that these phenomena brought the physicist to the explanation of magnetic phenomena through electrodynamic phenomena and, in this way, made the assumption of magnetic masses superfluous. This progress had, indeed, to be paid for by increasing the complexity of the forces of interaction which had to be assumed as existing between electrical masses in motion.

The escape from this unsatisfactory situation by the electric field theory of Faraday and Maxwell represents probably the most profound transformation of the foundations of physics since Newton's time. Again, it has been a step in the direction of constructive speculation which has increased the distance between the foundation of the theory and sense experiences. The existence of the field manifests itself, indeed, only when electrically charged bodies are introduced into it. The differential equations of Maxwell connect the spatial and temporal differential coefficients of the electric and magnetic fields. The electric masses are nothing more than places of non-vanishing divergence of the electric field. Light waves appear as undulatory electromagnetic field processes in space.

To be sure, Maxwell still tried to interpret his field theory mechanically by means of mechanical ether models. But these attempts receded gradually to the background following the representation of the theory—purged of any unnecessary trimmings—by Heinrich Hertz, so that in this theory the field finally took the fundamental position which had been occupied in Newton's mechanics by the material points. Primarily, however, this applied only for electromagnetic fields in empty space.

In its initial stage the theory was yet quite unsatisfactory for the interior of matter, because there, two electric vectors had to be introduced, which were connected by relations dependent on the nature of the medium, these relations being inaccessible to any theoretical analysis. An analogous situation arose in connection with the magnetic field, as well as in the relation between electric current density and the field.

Here H. A. Lorentz found a way out which showed, at the same time, the way to an electrodynamic theory of bodies in motion, a theory which was more or less free from arbitrary assumptions. His theory was built on the following fundamental hypotheses:

Everywhere (including the interior of ponderable bodies) the seat of the field is the empty space. The participation of matter in electromagnetic phenomena has its origin only in the fact that the elementary particles of matter carry unalterable electric charges, and, on this account, are subject on the one hand to the actions of ponderomotive forces and on the other hand possess the property of generating a field. The elementary particles obey Newton's law of motion for material points.

This is the basis on which H. A. Lorentz obtained his synthesis of Newton's mechanics and Maxwell's field theory. The weakness of this theory lies in the fact that it tried to determine the phenomena by a combination of partial differential equations (Maxwell's field equations for empty space) and total differential equations (equations of motion of points), which procedure was obviously unnatural. The inadequacy of this point of view manifested itself in the necessity of assuming finite dimensions for the particles in order to prevent the electromagnetic field existing at their surfaces from becoming infinitely large. The theory failed, moreover, to give any explanation concerning the tremendous forces which hold the electric charges on the individual particles. H. A. Lorentz accepted these weaknesses of his theory, which were well known to him, in order to explain the phenomena correctly at least in general outline.

Furthermore, there was one consideration which pointed beyond the frame of Lorentz's theory. In the environment of an electrically charged body there is a magnetic field which furnishes an (apparent) contribution to its inertia. Should it not be possible to explain the *total* inertia of the particles electromagnetically? It is clear that this problem could be worked out satisfactorily only if the particles could be interpreted as regular solutions of the electromagnetic partial differential equations. The Maxwell equations in their original form do not, however, allow such a description of particles, because their corresponding solutions contain a singularity. Theoretical physicists have tried for a long time, therefore, to reach the goal by a modification of Maxwell's equations. These attempts have, however, not been crowned with success. Thus it happened that the goal of erecting a pure electromagnetic field theory of matter remained unattained for the time being, although in principle no objection could be raised against the possibility of reaching such a goal. The lack of any systematic method leading to a solution discouraged further attempts in this direction. What appears certain to me, however, is that, in the foundations of any consistent field theory, the particle concept must not appear in addition to the field concept. The whole theory must be based solely on partial differential equations and their singularity-free solutions.

IV. THE THEORY OF RELATIVITY

There is no inductive method which could lead to the fundamental concepts of physics. Failure to understand this fact constituted the basic philosophical error of so many investigators of the nineteenth century. It was probably the reason why the molecular theory and Maxwell's theory were able to establish themselves only at a relatively late date. Logical thinking is necessarily deductive; it is based upon hypothetical concepts and axioms. How can we expect to choose the latter so that we might hope for a confirmation of the consequences derived from them?

The most satisfactory situation is evidently to be found in cases where the new fundamental hypotheses are suggested by the world of experience itself. The hypothesis of the non-existence of perpetual motion as a basis for thermodynamics affords such an example of a fundamental hypothesis suggested by experience; the same holds for Galileo's principle of inertia. In the same category, moreover, we find the fundamental hypotheses of the theory of relativity, which theory has led to an unexpected expansion and broadening of the field theory, and to the superseding of the foundations of classical mechanics.

The success of the Maxwell-Lorentz theory has given great confidence in the validity of the electromagnetic equations for empty space, and hence, in particular, in the assertion that light travels "in space" with a certain constant speed c. Is this assertion of the constancy of light velocity valid for every inertial system? If it were not, then one specific inertial system or, more accurately, one specific state of motion (of a body of reference) would be distinguished from all others. This, however, appeared to contradict all mechanical and electromagnetic-optical experimental facts.

For these reasons it was necessary to raise to the rank of a principle the validity of the law of constancy of light velocity for all inertial systems. From this, it follows that the spatial coordinates X_1, X_2, X_3, and the time X_4, must be transformed according to the "Lorentz-transformation" which is characterized by the invariance of the expression

$$ds^2 = dx_1{}^2 + dx_2{}^2 + dx_3{}^2 - dx_4{}^2$$

(if the unit of time is chosen in such a manner that the speed of light $c = 1$).

By this procedure time lost its absolute character, and was adjoined to the "spatial" coordinates as of algebraically (nearly) similar character. The absolute character of time and particularly of simultaneity was destroyed, and the four-dimensional description was introduced as the only adequate one.

In order to account, also, for the equivalence of all inertial systems with regard to all the phenomena of nature, it is necessary to postulate invariance of all systems of physical equations which express general laws with respect to Lorentz transformations. The elaboration of this requirement forms the content of the special theory of relativity.

This theory is compatible with the equations of Maxwell; but it is incompatible with the basis of classical mechanics. It is true that the equations of motion of the material point can be modified (and with them the expressions for momentum and kinetic energy of the material point) in such a manner as to satisfy the theory; but, the concept of the force of interaction, and with it the concept of potential energy of a system, lose their basis, because these concepts rest upon the idea of absolute simultaneity. The field, as determined by differential equations, takes the place of the force.

Since the foregoing theory allows interaction only by fields, it requires a field theory of gravitation. Indeed, it is not difficult to formulate such a theory in which, as in Newton's theory, the gravitational fields can be reduced to a scalar which is the solution of a partial differential equation. However, the experimental facts expressed in Newton's theory of gravitation lead in another direction, that of the general theory of relativity.

It is an unsatisfactory feature of classical mechanics that in its fundamental laws the same mass constant appears in two different rôles, namely as "inertial mass" in the law of motion, and as "gravitational mass" in the law of gravitation. As a result, the acceleration of a body in a pure gravitational field is

independent of its material; or, in a uniformly accelerated coordinate system (accelerated in relation to an "inertial system")
the motions take place as they would in a homogeneous gravitational field (in relation to a "motionless" system of coordinates).
If one assumes that the equivalence of these two cases is complete, then one attains an adaptation of our theoretical thinking
to the fact that the gravitational and inertial masses are equal.

From this it follows that there is no longer any reason for
favoring, as a matter of principle, the "inertial systems"; and,
we must admit on an equal footing also *non-linear* transformations of the coordinates (x_1, x_2, x_3, x_4). If we make such a
transformation of a system of coordinates of the special theory
of relativity, then the metric

$$ds^2 = dx_1{}^2 + dx_2{}^2 + dx_3{}^2 - dx_4{}^2$$

goes over into a general (Riemannian) metric of the form

$$ds^2 = g_{\mu\nu}\, dx_\mu\, dx_\nu \quad \text{(summed over } \mu \text{ and } \nu)$$

where the $g_{\mu\nu}$, symmetrical in μ and ν, are certain functions of
$x_1 \ldots x_4$ which describe both the metric properties, and the
gravitational field in relation to the new system of coordinates.

The foregoing improvement in the interpretation of the
mechanical basis must, however, be paid for in that—as becomes
evident on closer scrutiny—the new coordinates can no longer
be interpreted as results of measurements on rigid bodies and
clocks, as they could in the original system (an inertial system
with vanishing gravitational field).

The passage to the general theory of relativity is realized by
the assumption that such a representation of the field properties
of space already mentioned, by functions $g_{\mu\nu}$ (that is to say, by
a Riemann metric), is also justified in the *general* case in which
there is no system of coordinates in relation to which the metric
takes the simple quasi-Euclidean form of the special theory of
relativity.

Now the coordinates, by themselves, no longer express metric
relations, but only the "closeness" of objects whose coordinates
differ but little from one another. All transformations of the
coordinates have to be admitted so long as these transformations
are free from singularities. Only such equations as are covariant in relation to arbitrary transformations in this sense have

meaning as expressions of general laws of nature (postulate of general covariance).

The first aim of the general theory of relativity was a preliminary version which, while not meeting the requirements for constituting a closed system, could be connected in as simple a manner as possible with "directly observable facts." If the theory were restricted to pure gravitational mechanics, Newton's gravitational theory could serve as a model. This preliminary version may be characterized as follows:

1. The concept of the material point and of its mass is retained. A law of motion is given for it, this law of motion being the translation of the law of inertia into the language of the general theory of relativity. This law is a system of total differential equations, the system characteristic of the geodesic line.

2. Newton's law of interaction by gravitation is replaced by the system of the simplest generally covariant differential equations which can be set up for the $g_{\mu\nu}$-tensor. It is formed by equating to zero the once contracted Riemannian curvature tensor ($R_{\mu\nu} = 0$).

This formulation permits the treatment of the problem of the planets. More accurately speaking, it allows the treatment of the problem of motion of material points of practically negligible mass in the (centrally symmetric) gravitational field produced by a material point supposed to be "at rest." It does not take into account the reaction of the "moving" material points on the gravitational field, nor does it consider how the central mass produces this gravitational field.

Analogy with classical mechanics shows that the following is a way to complete the theory. One sets up as field equations

$$R_{ik} - \tfrac{1}{2}g_{ik}R = -T_{ik}$$

where R represents the scalar of Riemannian curvature, T_{ik} the energy tensor of the matter in a phenomenological representation. The left side of the equation is chosen in such a manner that its divergence disappears identically. The resulting disappearance of the divergence of the right side produces the "equations of motion" of matter, in the form of partial differential equations for the case where T_{ik} introduces, for the description of the matter, only *four* further independent functions (for instance, density, pressure, and velocity components, where there is between the latter an identity, and between pressure and density an equation of condition).

By this formulation one reduces the whole mechanics of gravitation to the solution of a single system of covariant partial differential equations. The theory avoids all the shortcomings which we have charged against the basis of classical mechanics. It is sufficient—as far as we know—for the representation of the observed facts of celestial mechanics. But it is similar to a building, one wing of which is made of fine marble (left part of the equation), but the other wing of which is built of low-grade wood (right side of equation). The phenomenological representation of matter is, in fact, only a crude substitute for a representation which would do justice to all known properties of matter.

There is no difficulty in connecting Maxwell's theory of the electromagnetic field with the theory of the gravitational field so long as one restricts himself to space free of ponderable matter and free of electric density. All that is necessary is to put on the right-hand side of the above equation for T_{ik} the energy tensor of the electromagnetic field in empty space and to adjoin to the so modified system of equations the Maxwell field equation for empty space, written in general covariant form. Under these conditions there will exist, between all these equations, a sufficient number of differential identities to guarantee their consistency. We may add that this necessary formal property of the total system of equations leaves arbitrary the choice of the sign of the member T_{ik}, a fact which later turned out to be important.

The desire to have, for the foundations of the theory, the greatest possible unity has resulted in several attempts to include the gravitational field and the electromagnetic field in one unified formal picture. Here we must mention particularly the five-dimensional theory of Kaluza and Klein. Having considered this possibility very carefully, I feel that it is more desirable to accept the lack of internal uniformity of the original theory, because I do not think that the totality of the hypotheses at the basis of the five-dimensional theory contains less arbitrary features than does the original theory. The same statement may be made for the projective version of the theory, which has been elaborated with great care, in particular, by v. Dantzig and by Pauli.

The foregoing considerations concern, exclusively, the theory of the field, free of matter. How are we to proceed from this point in order to obtain a complete theory of atomically constituted matter? In such a theory, singularities must certainly be excluded, since without such exclusion the differential equations do not completely determine the total field. Here, in the field theory of general relativity, we meet the same problem of a field-theoretical representation of matter as was met originally in connection with the pure Maxwell theory.

Here again the attempt of a field-theoretical construction of particles leads apparently to singularities. Here also the endeavor has been made to overcome this defect by the introduction of new field variables and by elaborating and extending the system of field equations. Recently, however, I discovered, in collaboration with Dr. Rosen, that the above-mentioned simplest combination of the field equations of gravitation and electricity produces centrally symmetrical solutions which can be represented as free of singularity (the well-known centrally symmetrical solutions of Schwarzschild for the pure gravitational field, and those of Reissner for the electric field with consideration of its gravitational action). We shall refer to this shortly in the paragraph next but one. In this way it seems possible to get for matter and its interactions a pure field theory free of additional hypotheses, one moreover whose test by submission to facts of experience does not lead to difficulties other than purely mathematical ones, which difficulties, however, are very serious.

V. QUANTUM THEORY AND THE FUNDAMENTALS OF PHYSICS

The theoretical physicists of our generation are expecting the erection of a new theoretical basis for physics which would make use of fundamental concepts greatly different from those of the field theory considered up to now. The reason is that it has been found necessary to use—for the mathematical representation of the so-called quantum phenomena—entirely new methods.

While the failure of classical mechanics, as revealed by the theory of relativity, is connected with the finite speed of light (its not being ∞), it was discovered at the beginning of our century that there were other kinds of inconsistencies between deductions from mechanics and experimental facts, which inconsistencies are connected with the finite magnitude (its not being zero) of Planck's constant h. In particular, while molecular mechanics requires that both heat content and (monochromatic) radiation density of solid bodies should decrease *in proportion* to the decreasing absolute temperature, experience has shown that they decrease much more rapidly than the absolute temperature. For a theoretical explanation of this behavior it was necessary to assume that the energy of a mechanical system cannot assume arbitrary values, but only certain discrete values whose mathematical expressions were always dependent upon Planck's constant h. Moreover, this conception was essential for the theory of the atom (Bohr's theory). For the transitions of these states into one another—with or without emission or absorption of radiation—no causal laws could be given, but only statistical ones; and a similar conclusion holds for the radioactive decay of atoms, which was carefully investigated about the same time. For more than two decades physicists tried vainly to find a uniform interpretation of this "quantum character" of systems and phenomena. Such an attempt was successful about ten years ago, through the agency of two entirely different theoretical methods of attack. We owe one of these to Heisenberg and Dirac, and the other to de Broglie and Schrödinger. The mathematical equivalence of the two methods was soon recognized by Schrödinger. I shall try here to sketch the line of thought of de Broglie and Schrödinger, which lies closer to the physicist's method of thinking, and shall accompany the description with certain general considerations.

The question is first: How can one assign a discrete succession of energy values H_σ to a system specified in the sense of classical mechanics (the energy function is a given function of the coordinates q_r and the corresponding momenta p_r)? Planck's constant h relates the frequency H_σ/h to the energy values H_σ. It is therefore sufficient to assign to the system a succession of discrete *frequency* values. This reminds us of the fact that in acoustics a series of discrete frequency values is coordinated to a linear partial differential equation (for given boundary conditions) namely, the sinusoidal periodic solutions. In corresponding manner, Schrödinger set himself the task of coordinating a partial differential equation for a scalar function ψ to the given energy function $\mathcal{E}(q_r, p_r)$, where the q_r and the time t are independent variables. In this he succeeded '(for a complex function ψ) in such a manner that the theoretical values of the energy H_σ, as required by the statistical theory, actually resulted in a satisfactory manner from the periodic solutions of the equation.

To be sure, it did not happen to be possible to associate a definite movement, in the sense of mechanics of material points, with a definite solution $\psi(q_r, t)$ of the Schrödinger equation. This means that the ψ function does not determine, at any rate *exactly*, the story of the q_r as functions of the time t. According to Born, however, an interpretation of the physical meaning of the ψ functions was shown to be possible in the following manner: $\psi\bar{\psi}$ (the square of the absolute value of the complex function ψ) is the probability density at the point under consideration in the configuration-space of the q_r, at the time t. It is therefore possible to characterize the content of the Schrödinger equation in a manner, easy to be understood, but not quite accurate, as follows: it determines how the probability density of a statistical ensemble of systems varies in the configuration-space with the time. Briefly: the Schrödinger equation determines the change of the function ψ of the q_r with time.

It must be mentioned that the results of this theory contain —as limiting values—the results of particle mechanics if the wave-lengths encountered in the solution of the Schrödinger problem are everywhere so small that the potential energy varies by a practically infinitely small amount for a distance of one wave-length in the configuration-space. Under these conditions the following can in fact be shown: We choose a region G_0 in the configuration-space which, although large (in every direction) in relation to the wave-length, is small in relation to the relevant dimensions of the configuration-space. Under these conditions it is possible to choose a function ψ for an initial time t_0 in such a manner that it vanishes outside the region G_0, and behaves, according to the Schrödinger equation, in such a manner that it retains this property—approximately at least— also for a later time, but with the region G_0 having passed at that time t into another region G. In this manner one can, with a certain degree of approximation, speak of the motion of the region G as a whole, and one can approximate this motion by the motion of a point in the configuration-space. This motion then coincides with the motion which is required by the equations of classical mechanics.

Experiments on interference made with particle rays have given a brilliant proof that the wave character of the phenomena of motion as assumed by the theory does, really, correspond to the facts. In addition to this, the theory succeeded, easily, in demonstrating the statistical laws of the transition of a system from one quantum state to another under the action of external forces, which, from the standpoint of classical mechanics, appears as a miracle. The external forces were here represented by small time dependent additions to the potential energy. Now, while in classical mechanics, such additions can produce only correspondingly small changes of the system, in the quantum mechanics they produce changes of any magnitude however large, but with correspondingly small probability, a consequence in perfect harmony with experience. Even an understanding of the laws of radioactive decay, at least in broad outline, was provided by the theory.

Probably never before has a theory been evolved which has given a key to the interpretation and calculation of such a heterogeneous group of phenomena of experience as has quantum theory. In spite of this, however, I believe that the theory is apt to beguile us into error in our search for a uniform basis for physics, because, in my belief, it is an *incomplete* representation of real things, although it is the only one which can be built out of the fundamental concepts of force and material points (quantum corrections to classical mechanics). The incompleteness of the representation leads necessarily to the statistical nature (incompleteness) of the laws. I will now give my reasons for this opinion.

I ask first: How far does the ψ function describe a real state of a mechanical system? Let us assume the ψ_r to be the periodic solutions (put in the order of increasing energy values) of the Schrödinger equation. I shall leave open, for the time being, the question as to how far the individual ψ_r are *complete* descriptions of physical states. A system is first in the state ψ_1 of lowest energy $\mathcal{E}_1$. Then during a finite time a small disturbing force acts upon the system. At a later instant one obtains then from the Schrödinger equation a ψ function of the form

$$\psi = \Sigma \, c_r \psi_r$$

where the c_r are (complex) constants. If the ψ_r are "normalized," then $|c_1|$ is nearly equal to 1, $|c_2|$ etc. is small compared with 1. One may now ask: Does ψ describe a real state of the system? If the answer is yes, then we can hardly do otherwise than ascribe * to this state a definite energy $\mathcal{E}$, and, in particular, an energy which exceeds $\mathcal{E}_1$ by a small amount (in any case $\mathcal{E}_1 < \mathcal{E} < \mathcal{E}_2$). Such an assumption is, however, at variance with the experiments on electron impact such as have been made by J. Franck and G. Hertz, if one takes into account Millikan's demonstration of the discrete nature of electricity. As a matter of fact, these experiments lead to the conclusion that energy values lying between the quantum values do not exist. From this it follows that our function ψ does not in any way describe a homogeneous state of the system, but represents rather a statistical description in which the c_r represent probabilities of the individual energy values. It seems to be clear, therefore, that Born's statistical interpretation of quantum theory is the only possible one. The ψ function does not in any way describe a state which could be that of a single system; it relates rather to many systems, to "an en-

---

* Because, according to a well-established consequence of the relativity theory, the energy of a complete system (at rest) is equal to its inertia (as a whole). This, however, must have a well-defined value.

semble of systems" in the sense of statistical mechanics. If, except for certain special cases, the ψ function furnishes only *statistical* data concerning measurable magnitudes, the reason lies not only in the fact that the *operation of measuring* introduces unknown elements, which can be grasped only statistically, but because of the very fact that the ψ function does not, in any sense, describe the state of *one* single system. The Schrödinger equation determines the time variations which are experienced by the ensemble of systems which may exist with or without external action on the single system.

Such an interpretation eliminates also the paradox recently demonstrated by myself and two collaborators, and which relates to the following problem.

Consider a mechanical system consisting of two partial systems A and B which interact with each other only during a limited time. Let the ψ function before their interaction be given. Then the Schrödinger equation will furnish the ψ function after the interaction has taken place. Let us now determine the physical state of the partial system A as completely as possible by measurements. Then quantum mechanics allows us to determine the ψ function of the partial system B from the measurements made, and from the ψ function of the total system. This determination, however, gives a result which depends upon *which* of the physical quantities (observables) of A have been measured (for instance, coordinates *or* momenta). Since there can be only *one* physical state of B after the interaction which cannot reasonably be considered to depend on the particular measurement we perform on the system A separated from B it may be concluded that the ψ function is *not* unambiguously coordinated to the physical state. This coordination of several ψ functions to the same physical state of system B shows again that the ψ function cannot be interpreted as a (complete) description of a physical state of a single system. Here also the coordination of the ψ function to an ensemble of systems eliminates every difficulty.*

---

* A measurement on A, for example, thus involves a transition to a narrower ensemble of systems. The latter (hence also its ψ function) depends upon the point of view according to which this reduction of the ensemble of systems is carried out.

The fact that quantum mechanics affords, in such a simple manner, statements concerning (apparently) discontinuous transitions from one state to another without actually giving a description of the specific process—this fact is connected with another, namely, the fact that the theory, in reality, does not operate with the single system, but with a totality of systems. The coefficients c_r of our first example are really altered very little under the action of the external force. With this interpretation of quantum mechanics one can understand why this theory can easily account for the fact that weak disturbing forces are able to produce changes of any magnitude in the physical state of a system. Such disturbing forces produce, indeed, only correspondingly small changes of the *statistical density* in the ensemble of systems, and hence only infinitely weak changes of the ψ functions, the mathematical description of which offers far less difficulty than would be involved in the mathematical description of finite changes experienced by part of the single systems. What happens to the single system remains, it is true, entirely unclarified by this mode of consideration; this enigmatic event is entirely eliminated from the description by the statistical approach.

But now I ask: Is there really any physicist who believes that we shall never get any insight into these important changes in the single systems, in their structure and their causal connections, regardless of the fact that these single events have been brought so close to us, thanks to the marvelous inventions of the Wilson chamber and the Geiger counter? To believe this is logically possible without contradiction; but, it is so very contrary to my scientific instinct that I cannot forego the search for a more complete conception.

To these considerations we should add those of another kind which also appear to indicate that the methods introduced by quantum mechanics are not likely to give a useful basis for the whole of physics. In the Schrödinger equation, absolute time, and also the potential energy, play a decisive rôle, while these two concepts have been recognized by the theory of relativity as inadmissible in principle. If one wishes to escape from this difficulty, he must found the theory upon field and field laws instead of upon forces of interaction. This leads us to apply the statistical methods of quantum mechanics to fields, that is, to systems of infinitely many degrees of freedom. Although the attempts so far made are restricted to linear equations, which, as we know from the results of the general theory of relativity, are insufficient, the complications met up to now by the very ingenious attempts are already terrifying. They certainly will multiply if one wishes to obey the requirements of the general theory of relativity, the justification of which in principle nobody doubts.

To be sure, it has been pointed out that the introduction of a space-time continuum may be considered as contrary to nature in view of the molecular structure of everything which happens on a small scale. It is maintained that perhaps the success of the Heisenberg method points to a purely algebraical method of description of nature, that is, to the elimination of continuous functions from physics. Then, however, we must also give up, on principle, the space-time continuum. It is conceivable that human ingenuity will some day find methods which will make it possible to proceed along such a path. At the present time, however, such a program looks like an attempt to breathe in empty space.

There is no doubt that quantum mechanics has seized hold of a good deal of truth, and that it will be a touchstone for any future theoretical basis, in that it must be deducible as a limiting case from that basis, just as electrostatics is deducible from the Maxwell equations of the electromagnetic field or as thermodynamics is deducible from classical mechanics. However, I do not believe that quantum mechanics can serve as a *starting point* in the search for this basis, just as, vice versa, one could not find from thermodynamics (resp. statistical mechanics) the foundations of mechanics.

In view of this situation, it seems to be entirely justifiable seriously to consider the question as to whether the basis of field physics cannot by *any* means be put into harmony with quantum phenomena. Is this not the only basis which, with the presently available mathematical tools, can be adapted to the requirements of the general theory of relativity? The belief, prevailing among the physicists of today, that such an attempt would be hopeless, may have its root in the unwarranted assumption that such a theory must lead, in first approximation, to the equations of classical mechanics for the motion of corpuscles, or at least to total differential equations. As a matter of fact, up to now we have never succeeded in a field-theoretical description of corpuscles free of singularities, and we can, *a priori*, say nothing about the behavior of such entities. *One thing*, however, is certain: if a field theory results in a representation of corpuscles free of singularities, then the behavior of these corpuscles in time is determined solely by the differential equations of the field.

VI. Relativity Theory and Corpuscles

I shall now show that, according to the general theory of relativity, there exist singularity-free solutions of field equations which can be interpreted as representing corpuscles. I restrict myself here to neutral particles because, in another recent publication in collaboration with Dr. Rosen, I have treated this question in detail, and because the essentials of the problem can be completely exhibited in this case.

The gravitational field is entirely described by the tensor $g_{\mu\nu}$. In the three-index symbols $\Gamma^{\sigma}_{\mu\nu}$, there appear also the contravariant $g^{\mu\nu}$ which are defined as the minors of the $g_{\mu\nu}$ divided by the determinant $g(=|g_{\alpha\beta}|)$. In order that the R_{ik} shall be defined and finite, it is not sufficient that there shall be, in the neighborhood of every point of the continuum, a system of coordinates in which the $g_{\mu\nu}$ and their first differential quotients are continuous and differentiable, but it is also necessary that the determinant g shall nowhere vanish. This last restriction disappears, however, if one replaces the differential equations $R_{ik} = 0$ by $g^2 R_{ik} = 0$, the left-hand sides of which are *whole* rational functions of the g_{ik} and of their derivatives.

These equations have the centrally symmetrical solution given by Schwarzschild

$$ds^2 = -\frac{1}{1-2m/r}\, dr^2 - r^2(d\theta^2 + \sin^2\theta d\varphi^2) + \left(1 - \frac{2m}{r}\right) dt^2$$

This solution has a singularity at $r = 2m$, since the coefficient of dr^2 (i.e., g_{11}), becomes infinite on this hypersurface. If, how-ever, we replace the variable r by ρ defined by the equation

$$\rho^2 = r - 2m$$

we obtain

$$ds^2 = -4(2m + \rho^2)d\rho^2 - (2m + \rho^2)^2(d\theta^2 + \sin^2\theta d\varphi^2)$$

$$+ \frac{\rho^2}{2m + \rho^2} dt^2$$

This solution behaves regularly for all values of ρ. The vanish-ing of the coefficient of dt^2 (i.e., g_{44}) for $\rho = 0$ results, it is true, in the consequence that the determinant g vanishes for this value; but, with the methods of writing the field equations actually adopted, this does not constitute a singularity.

If ρ varies from $-\infty$ to $+\infty$, then r varies from $+\infty$ to $r = 2m$ and then back to $+\infty$, while for such values of r as correspond to $r < 2m$ there are no corresponding real values of ρ. Hence the Schwarzschild solution becomes a regular solution by representing the physical space as consisting of two iden-tical "sheets" in contact along the hypersurface $\rho = 0$ (i.e., $r = 2m$), on which the determinant g vanishes. Let us call such a connection between the two (identical) sheets a "bridge." Hence the existence of such a bridge between the two sheets in the finite realm corresponds to the existence of a material neutral particle which is described in a manner free from singularities.

The solution of the problem of the motion of neutral par-ticles evidently amounts to the discovery of such solutions of the gravitational equations (written free of denominators), as contain several bridges.

The conception sketched above corresponds, *a priori*, to the atomistic structure of matter in so far as the "bridge" is by its nature a discrete element. Moreover, we see that the mass constant m of the neutral particles must necessarily be positive, since no solution free of singularities can correspond to the Schwarzschild solution for a negative value of m. Only the examination of the several-bridge-problem can show whether or not this theoretical method furnishes an explanation of the empirically demonstrated equality of the masses of the particles found in nature, and whether it takes into account the facts which the quantum mechanics has so wonderfully comprehended.

In an analogous manner, it is possible to demonstrate that the combined equations of gravitation and electricity (with appropriate choice of the sign of the electrical member in the gravitational equations) produce a singularity-free bridge-representation of the electric corpuscle. The simplest solution of this kind is that for an electrical particle without gravitational mass.

So long as the considerable mathematical difficulties concerned with the solution of the several-bridge-problem are not overcome, nothing can be said concerning the usefulness of the theory from the physicist's point of view. However, it constitutes, as a matter of fact, the first attempt toward the consistent elaboration of a field theory which presents a possibility of explaining the properties of matter. In favor of this attempt one should also add that it is based on the simplest possible relativistic field equations known today.

SUMMARY

Physics constitutes a logical system of thought which is in a state of evolution, whose basis cannot be distilled, as it were, from experience by an inductive method, but can only be arrived at by free invention. The justification (truth content) of the system rests in the verification of the derived propositions by sense experiences, whereby the relations of the latter to the former can only be comprehended intuitively. Evolution is proceeding in the direction of increasing simplicity of the logical basis. In order further to approach this goal, we must resign to the fact that the logical basis departs more and more from the facts of experience, and that the path of our thought from the fundamental basis to those derived propositions, which correlate with sense experiences, becomes continually harder and longer.

Our aim has been to sketch, as briefly as possible, the development of the fundamental concepts in their dependence upon the facts of experience and upon the endeavor to achieve internal perfection of the system. These considerations were intended to illuminate the present state of affairs, as it appears to me. (It is unavoidable that a schematic historic exposition is subjectively colored.)

I try to demonstrate how the concepts of bodily objects, space, subjective and objective time, are connected with one another and with the nature of our experience. In classical mechanics the concepts of space and time become independent. The concept of the bodily object is replaced in the foundations by the concept of the material point, by which means mechanics becomes fundamentally atomistic. Light and electricity produce insurmountable difficulties when one attempts to make mechanics the basis of all physics. We are thus led to the field theory of electricity, and, later on to the attempt to base physics entirely upon the concept of the field (after an attempted compromise with classical mechanics). This attempt leads to the theory of relativity (evolution of the notion of space and time into that of the continuum with metric structure).

I try to demonstrate, furthermore, why in my opinion quantum theory does not seem capable to furnish an adequate foundation for physics: one becomes involved in contradictions if one tries to consider the theoretical quantum description as a *complete* description of the individual physical system or event.

On the other hand, the field theory is as yet unable to explain the molecular structure of matter and of quantum phenomena. It is shown, however, that the conviction of the inability of field theory to solve these problems by its methods rests upon prejudice.

THE FUNDAMENTS OF THEORETICAL PHYSICS

From Science, *Washington, D. C. May 24, 1940.*

Science is the attempt to make the chaotic diversity of our sense-experience correspond' to a logically uniform system of thought. In this system single experiences must be correlated with the theoretic structure in such a way that the resulting coordination is unique and convincing.

The sense-experiences are the given subject-matter. But the theory that shall interpret them is man-made. It is the result of an extremely laborious process of adaptation: hypothetical, never completely final, always subject to question and doubt.

The scientific way of forming concepts differs from that which we use in our daily life, not basically, but merely in the more precise definition of concepts and conclusions; more painstaking and systematic choice of experimental material; and greater logical economy. By this last we mean the effort to reduce all concepts and correlations to as few as possible logically independent basic concepts and axioms.

What we call physics comprises that group of natural sciences which base their concepts on measurements; and whose concepts and propositions lend themselves to mathematical formulation. Its realm is accordingly defined as that part of the sum total of our knowledge which is capable of being expressed in mathematical terms. With the progress of science, the realm of physics has so expanded that it seems to be limited only by the limitations of the method itself.

The larger part of physical research is devoted to the development of the various branches of physics, in each of which the object is the theoretical understanding of more or less restricted fields of experience, and in each of which the laws and concepts remain as closely as possible related to experience. It is this department of science, with its ever-growing specialization, which has revolutionized practical life in the last centuries, and given birth to the possibility that man may at last be freed from the burden of physical toil.

On the other hand, from the very beginning there has always been present the attempt to find a unifying theoretical basis for all these single sciences, consisting of a minimum of concepts and fundamental relationships, from which all the concepts and relationships of the single disciplines might be derived by logical process. This is what we mean by the search for a foundation of the whole of physics. The confident belief that this ultimate goal may be reached is the chief source of the passionate devotion which has always animated the researcher. It is in this sense that the following observations are devoted to the foundations of physics.

From what has been said it is clear that the word foundations in this connection does not mean something analogous in all respects to the foundations of a building. Logically considered, of course, the various single laws of physics rest upon this foundation. But whereas a building may be seriously damaged by a heavy storm or spring flood, yet its foundations remain intact, in science the logical foundation is always in greater peril from new experiences or new knowledge than are the branch disciplines with their closer experimental contacts. In the connection of the foundation with all the single parts lies its great significance, but likewise its greatest danger in face of any new factor. When we realize this, we are led to wonder why the so-called revolutionary epochs of the science of physics have not more often and more completely changed its foundation than has actually been the case.

The first attempt to lay a uniform theoretical foundation was the work of Newton. In his system everything is reduced to the following concepts: (1) Mass points with invariable mass; (2) action at a distance between any pair of mass points; (3) law of motion for the mass point. There was not, strictly speaking, any all-embracing foundation, because an explicit law was formulated only for the actions-at-a-distance of gravitation; while for other actions-at-a-distance nothing was established *a priori* except the law of equality of *actio* and *reactio*. Moreover, Newton himself fully realized that time and space were essential elements, as physically effective factors, of his system, if only by implication.

This Newtonian basis proved eminently fruitful and was regarded as final up to the end of the nineteenth century. It not only gave results for the movements of the heavenly bodies, down to the most minute details, but also furnished a theory of the mechanics of discrete and continuous masses, a simple explanation of the principle of the conservation of energy and a complete and brilliant theory of heat. The explanation of the facts of electrodynamics within the Newtonian system was more forced; least convincing of all, from the very beginning, was the theory of light.

It is not surprising that Newton would not listen to a wave theory of light; for such a theory was most unsuited to his theoretical foundation. The assumption that space was filled with a medium consisting of material points that propagated light waves without exhibiting any other mechanical properties must have seemed to him quite artificial. The strongest empirical arguments for the wave nature of light, fixed speeds of propagation, interference, diffraction, polarization were either unknown or else not known in any well-ordered synthesis. He was justified in sticking to his corpuscular theory of light.

During the nineteenth century the dispute was settled in favor of the wave theory. Yet no serious doubt of the mechanical foundation of physics arose, in the first place because nobody knew where to find a foundation of another sort. Only slowly, under the irresistible pressure of facts, there developed a new foundation of physics, field-physics.

From Newton's time on, the theory of action-at-a-distance was constantly found artificial. Efforts were not lacking to explain gravitation by a kinetic theory, that is, on the basis of collision forces of hypothetical mass particles. But the attempts were superficial and bore no fruit. The strange part played by space (or the inertial system) within the mechanical foundation was also clearly recognized, and criticized with especial clarity by Ernst Mach.

The great change was brought about by Faraday, Maxwell, and Hertz—as a matter of fact half-unconsciously and against their will. All three of them, throughout their lives, considered themselves adherents of the mechanical theory. Hertz had found the simplest form of the equations of the electromagnetic field, and declared that any theory leading to these equations was Maxwellian theory. Yet toward the end of his short life he wrote a paper in which he presented as the foundation of physics a mechanical theory freed from the force-concept.

For us, who took in Faraday's ideas so to speak with our mother's milk, it is hard to appreciate their greatness and audacity. Faraday must have grasped with unerring instinct the artificial nature of all attempts to refer electromagnetic phenomena to actions-at-a-distance between electric particles reacting on each other. How was each single iron filing among a lot scattered on a piece of paper to know of the single electric particles running round in a nearby conductor? All these electric particles together seemed to create in the surrounding space a condition which in turn produced a certain order in the filings. These spatial states, today called fields, if their geometrical structure and interdependent action were once rightly grasped, would, he was convinced, furnish the clue to the mysterious electromagnetic interactions. He conceived these fields as states of mechanical stress in a space-filling medium, similar to the states of stress in an elastically distended body. For at that time this was the only way one could conceive of states that were apparently continuously distributed in space. The peculiar type of mechanical interpretation of these fields remained in the background—a sort of placation of the scientific conscience in view of the mechanical tradition of Faraday's time. With the help of these new field concepts Faraday succeeded in forming a qualitative concept of the whole complex of electromagnetic effects discovered by him and his predecessors. The precise formulation of the time-space laws of those fields was the work of Maxwell. Imagine his feelings when the differential equations he had formulated proved to him that electromagnetic fields spread in the form of polarized waves and with the speed of light! To few men in the world has such an experience been vouchsafed. At that thrilling moment he surely never guessed that the riddling nature of light, apparently so completely solved, would continue to baffle succeeding generations. Meantime, it took physicists some decades to grasp the full significance of Maxwell's discovery, so bold was the leap that his genius forced upon the conceptions of his fellow-workers. Only after Hertz had demonstrated experimentally the existence of Maxwell's electromagnetic waves did resistance to the new theory break down.

But if the electromagnetic field could exist as a wave independent of the material source, then the electrostatic interaction could no longer be explained as action-at-a-distance. And what was true for electrical action could not be denied for gravitation. Everywhere Newton's actions-at-a-distance gave way to fields spreading with finite velocity.

Of Newton's foundation there now remained only the maerial mass points subject to the law of motion. But J. J. Thomson pointed out that an electrically charged body in motion must, according to Maxwell's theory, possess a magnetic field whose energy acted precisely as does an increase of kinetic energy to the body. If, then, a part of kinetic energy consists of field energy, might that not then be true of the whole of the kinetic energy? Perhaps the basic property of matter, its inertia, could be explained within the field theory? The question led to the problem of an interpretation of matter in terms of field theory, the solution of which would furnish an explanation of the atomic structure of matter. It was soon realized that Maxwell's theory could not accomplish such a program. Since then many scientists have zealously sought to complete the field theory by some generalization that should comprise a theory of matter; but so far such efforts have not been crowned with success. In order to construct a theory, it is not enough to have a clear conception of the goal. One must also have a formal point of view which will sufficiently restrict the unlimited variety of possibilities. So far this has not been found; accordingly the field theory has not succeeded in furnishing a foundation for the whole of physics.

For several decades most physicists clung to the conviction that a mechanical substructure would be found for Maxwell's theory. But the unsatisfactory results of their efforts led to gradual acceptance of the new field concepts as irreducible fundamentals—in other words, physicists resigned themselves to giving up the idea of a mechanical foundation.

Thus physicists held to a field-theory program. But it could not be called a foundation, since nobody could tell whether a consistent field theory could ever explain on the one hand gravitation, on the other hand the elementary components of matter. In this state of affairs it was necessary to think of material particles as mass points subject to Newton's laws of motion. This was the procedure of Lorentz in creating his electron theory and the theory of the electromagnetic phenomena of moving bodies.

Such was the point at which fundamental conceptions had arrived at the turn of the century. Immense progress was made in the theoretical penetration and understanding of whole groups of new phenomena; but the establishment of a unified foundation for physics seemed remote indeed. And this state of things has even been aggravated by subsequent developments. The development during the present century is characterized by two theoretical systems essentially independent of each other: the theory of relativity and the quantum theory. The two systems do not directly contradict each other; but they seem little adapted to fusion into one unified theory. We must briefly discuss the basic idea of these two systems.

The theory of relativity arose out of efforts to improve, with reference to logical economy, the foundation of physics as it existed at the turn of the century. The so-called special or restricted relativity theory is based on the fact that Maxwell's equations (and thus the law of propagation of light in empty space) are converted into equations of the same form, when they undergo Lorentz transformation. This formal property of the Maxwell equations is supplemented by our fairly secure empirical knowledge that the laws of physics are the same with respect to all inertial systems. This leads to the result that the Lorentz transformation—applied to space and time coordinates—must govern the transition from one inertial system to any other. The content of the restricted relativity theory can accordingly be summarized in one sentence: all natural laws must be so conditioned that they are covariant with respect to Lorentz transformations. From this it follows that the simultaneity of two distant events is not an invariant concept and that the dimensions of rigid bodies and the speed of clocks depend upon their state of motion. A further consequence was a modification of Newton's law of motion in cases where the speed of a given body was not small compared with the speed of light. There followed also the principle of the equivalence of mass and energy, with the laws of conservation of mass and energy becoming one and the same. Once it was shown that simultaneity was relative and depended on the frame of reference, every possibility of retaining actions-at-a-distance within the foundation of physics disappeared, since that concept presupposed the absolute character of simultaneity (it must be possible to state the location of the two interacting mass points "at the same time").

The general theory of relativity owes its origin to the attempt to explain a fact known since Galileo's and Newton's time but hitherto eluding all theoretical interpretation: the inertia and the weight of a body, in themselves two entirely distinct things, are measured by one and the same constant, the mass. From this correspondence follows that it is impossible to discover by experiment whether a given system of coordinates is accelerated, or whether its motion is straight and uniform and the observed effects are due to a gravitational field (this is the equivalence principle of the general relativity theory). It shatters the concepts of the inertial system, as soon as gravitation enters in. It may be remarked here that the inertial system is a weak point of the Galilean-Newtonian mechanics. For there is presupposed a mysterious property of physical space, conditioning the kind of coordinate-systems for which the law of inertia and the Newtonian law of motion hold good.

These difficulties can be avoided by the following postulate: natural laws are to be formulated in such a way that their form is identical for coordinate systems of any kind of states of motion. To accomplish this is the task of the general theory of relativity. On the other hand, we deduce from the restricted theory the existence of a Riemannian metric within the time-space continuum, which, according to the equivalence principle, describes both the gravitational field and the metric properties of space. Assuming that the field equations of gravitation are of the second differential order, the field law is clearly determined.

Aside from this result, the theory frees field physics from the disability it suffered from, in common with the Newtonian mechanics, of ascribing to space those independent physical properties which heretofore had been concealed by the use of an inertial system. But it cannot be claimed that those parts of the general relativity theory which can today be regarded as final have furnished physics with a complete and satisfactory foundation. In the first place, the total field appears in it to be composed of two logically unconnected parts, the gravitational and the electromagnetic. And in the second place, this theory, like the earlier field theories, has not up till now supplied an explanation of the atomistic structure of matter. This failure has probably some connection with the fact that so far it has contributed nothing to the understanding of quantum phenomena. To take in these phenomena, physicists have been driven to the adoption of entirely new methods, the basic characteristics of which we shall now discuss.

In the year nineteen hundred, in the course of a purely theoretic investigation, Max Planck made a very remarkable discovery: the law of radiation of bodies as a function of temperature could not be derived solely from the laws of Maxwellian electrodynamics. To arrive at results consistent with the relevant experiments, radiation of a given frequency had to be treated as though it consisted of energy atoms of the individual energy $h\nu$, where h is Planck's universal constant. During the years following, it was shown that light was everywhere produced and absorbed in such energy quanta. In particular Niels Bohr was able largely to understand the structure of the atom, on the assumption that atoms can have only discrete energy values, and that the discontinuous transitions between them are connected with the emission or absorption of such an energy quantum. This threw some light on the fact that in their gaseous state elements and their compounds radiate and absorb only light of certain sharply defined frequencies. All this was quite inexplicable within the frame of the hitherto existing theories. It was clear that at least in the field of atomistic phenomena the character of everything that happens is determined by discrete states and by apparently discontinuous transitions between them, Planck's constant h playing a decisive role.

The next step was taken by de Broglie. He asked himself how the discrete states could be understood by the aid of the current concepts, and hit on a parallel with stationary waves, as for instance in the case of the proper frequencies of organ pipes and strings in acoustics. True, wave actions of the kind here required were unknown; but they could be constructed, and their mathematical laws fomulated, employing Planck's constant h. De Broglie conceived an electron revolving about the atomic nucleus as being connected with such a hypothetical wave train, and made intelligible to some extent the discrete character of Bohr's "permitted" paths by the stationary character of the corresponding waves.

Now in mechanics the motion of material points is determined by the forces or fields of force acting upon them. Hence it was to be expected that those fields of force would also influence de Broglie's wave fields in an analogous way. Erwin Schrödinger showed how this influence was to be taken into account, re-interpreting by an ingenious method certain formulations of classical mechanics. He even succeeded in expanding the wave mechanical theory to a point where without the introduction of any additional hypotheses, it became applicable to any mechanical system consisting of an arbitrary number of mass points, that is to say possessing an arbitrary number of degrees of freedom. This was possible because a mechanical system consisting of n mass points is mathematically equivalent to a considerable degree to one single mass point moving in a space of $3\ n$ dimensions.

On the basis of this theory there was obtained a surprisingly good representation of an immense variety of facts which otherwise appeared entirely incomprehensible. But on one point, curiously enough, there was failure: it proved impossible to associate with these Schrödinger waves definite motions of the mass points—and that, after all, had been the original purpose of the whole construction.

The difficulty appeared insurmountable, until it was overcome by Born in a way as simple as it was unexpected. The de Broglie-Schrödinger wave fields were not to be interpreted as a mathematical description of how an event actually takes place in time and space, though, of course, they have reference to such an event. Rather they are a mathematical description of what we can actually know about the system. They serve only to make statistical statements and predictions of the results of all measurements which we can carry out upon the system.

Let me illustrate these general features of quantum mechanics by means of a simple example: we shall consider a mass point kept inside a restricted region G by forces of finite strength. If the kinetic energy of the mass point is below a certain limit, then the mass point, according to classical mechanics, can never leave the region G. But according to quantum mechanics, the mass point, after a period not immediately predictable, is able to leave the region G, in an unpredictable direction, and escape into surrounding space. This case, according to Gamow, is a simplified model of radioactive disintegration.

The quantum theoretical treatment of this case is as follows: at the time t_0 we have a Schrödinger wave system entirely inside G. But from the time t_0 onwards, the waves leave the interior of G in all directions, in such a way that the amplitude of the outgoing wave is small compared to the initial amplitude of the wave system inside G. The further these outside waves spread, the more the amplitude of the waves inside G diminishes, and correspondingly the intensity of the later waves issuing from G. Only after infinite time has passed is the wave supply inside G exhausted, while the outside wave has spread over an ever-increasing space.

But what has this wave process to do with the first object of our interest, the particle originally enclosed in G? To answer this question, we must imagine some arrangement which will permit us to carry out measurements on the particle. For instance, let us imagine somewhere in the surrounding space a screen so made that the particle sticks to it on coming into contact with it. Then, from the intensity of the waves hitting the screen at some point, we draw conclusions as to the probability of the particle hitting the screen there at that time. As soon as the particle has hit any particular point of the screen, the whole wave field loses all its physical meaning; its only purpose was to make probability predictions as to the place and time of the particle hitting the screen (or, for instance, its momentum at the time when it hits the screen).

All other cases are analogous. The aim of the theory is to determine the probability of the results of measurement upon a system at a given time. On the other hand, it makes no attempt to give a mathematical representation of what is actually present or goes on in space and time. On this point the quan-

tum theory of today differs fundamentally from all previous
theories of physics, mechanistic as well as field theories. Instead
of a model description of actual space-time events, it gives the
probability distributions for possible measurements as functions
of time.

It must be admitted that the new theoretical conception owes
its origin not to any flight of fancy but to the compelling force
of the facts of experience. All attempts to represent the particle
and wave features displayed in the phenomena of light and
matter, by direct recourse to a. space-time model, have so far
ended in failure. And Heisenberg has convincingly shown,
from an empirical point of view, that any decision as to a rigor-
ously deterministic structure of nature is definitely ruled out, because of the atomistic structure of our experimental apparatus.
Thus it is probably out of the question that any future knowledge can compel physics again to relinquish our present statistical theoretical foundation in favor of a deterministic one
which would deal directly with physical reality. Logically the
problem seems to offer two possibilities, between which we
are in principle given a choice. In the end the choice will be
made according to which kind of description yields the formulation of the simplest foundation, logically speaking. At the
present, we are quite without any deterministic theory directly
describing the events themselves and in consonance with the
facts.

For the time being, we have to admit that we do not possess
any general theoretical basis for physics, which can be regarded
as its logical foundation. The field theory, so far, has failed
in the molecular sphere. It is agreed on all hands that the only
principle which could serve as the basis of quantum theory
would be one that constituted a translation of the field theory
into the scheme of quantum statistics. Whether this will
actually come about in a satisfactory manner, nobody can
venture to say.

Some physicists, among them myself, cannot believe that we
must abandon, actually and forever, the idea of direct representation of physical reality in space and time; or that we must
accept the view that events in nature are analogous to a game

of chance. It is open to every man to choose the direction of his striving; and also every man may draw comfort from Lessing's fine saying, that the search for truth is more precious than its possession.

THE COMMON LANGUAGE OF SCIENCE

Broadcast recording for Science Conference, London, September 28, 1941. Published in Advancement of Science, *London, Vol. 2, No. 5.*

The first step toward language was to link acoustically or otherwise commutable signs to sense-impressions. Most likely all sociable animals have arrived at this primitive kind of communication—at least to a certain degree. A higher development is reached when further signs are introduced and understood which establish relations between those other signs designating sense-impression. At this stage it is already possible to report somewhat complex series of impressions; we can say that language has come to existence. If language is to lead at all to understanding, there must be rules concerning the relations between the signs on the one hand, and on the other hand there must be a stable correspondence between signs and impressions. In their childhood individuals connected by the same language grasp these rules and relations mainly by intuition. When man becomes conscious of the rules concerning the relations between signs, the so-called grammar of language is established.

In an early stage the words may correspond directly to impressions. At a later stage this direct connection is lost in so far as some words convey relations to perceptions only if used in connection with other words (for instance such words as: "is," "or," "thing"). Then word-groups rather than single words refer to perceptions. When language becomes thus partially independent from the background of impressions a greater inner coherence is gained.

Only at this further development where frequent use is made of so-called abstract concepts, language becomes an instrument of reasoning in the true sense of the word. But it is also this development which turns language into a dangerous source of error and deception. Everything depends on the degree to which words and word-combinations correspond to the world of impression.

What is it that brings about such an intimate connection between language and thinking? Is there no thinking without the use of language, namely in concepts and concept-combinations for which words need not necessarily come to mind? Has not every one of us struggled for words although the connection between "things" was already clear?

We might be inclined to attribute to the act of thinking complete independence from language if the individual formed or were able to form his concepts without the verbal guidance of his environment. Yet most likely the mental shape of an individual, growing up under such conditions, would be very poor. Thus we may conclude that the mental development of the individual and his way of forming concepts depend to a high degree upon language. This makes us realize to what extent the same language means the same mentality. In this sense thinking and language are linked together.

What distinguishes the language of science from language as we ordinarily understand the word? How is it that scientific language is international? What science strives for is an utmost acuteness and clarity of concepts as regards their mutual relation and their correspondence to sensory data. As an illustration let us take the language of Euclidean geometry and algebra. They manipulate with a small number of independently introduced concepts, respectively symbols, such as the integral number, the straight line, the point, as well as with signs which designate the fundamental operations, that is, the connections between those fundamental concepts. This is the basis for the construction, and respectively the definition of all other statements and concepts. The connection between concepts and statements on the one hand and the sensory data on the other hand is established through acts of counting and measuring whose performance is sufficiently well determined.

The supernational character of scientific concepts and scientific language is due to the fact that they have been set up by the best brains of all countries and all times. In solitude, and yet in cooperative effort as regards the final effect, they created the spiritual tools for the technical revolutions which have transformed the life of mankind in the last centuries. Their system of concepts has served as a guide in the bewildering chaos of perceptions so that we learned to grasp general truths from particular observations.

What hopes and fears does the scientific method imply for mankind? I do not think that this is the right way to put the question. Whatever this tool in the hand of man will produce depends entirely on the nature of the goals alive in this mankind. Once these goals exist, the scientific method furnishes means to realize them. Yet it cannot furnish the very goals. The scientific method itself would not have led anywhere, it would not even have been born without a passionate striving for clear understanding.

Perfection of means and confusion of goals seem—in my opinion—to characterize our age. If we desire sincerely and passionately the safety, the welfare, and the free development of the talents of all men, we shall not be in want of the means to approach such a state. Even if only a small part of mankind strives for such goals, their superiority will prove itself in the long run.

E = M C²

From Science Illustrated, *New York, April, 1946.*

In order to understand the law of the equivalence of mass and energy, we must go back to two conservation or "balance" principles which, independent of each other, held a high place in pre-relativity physics. These were the principle of the conservation of energy and the principle of the conservation of mass. The first of these, advanced by Leibnitz as long ago as the seventeenth century, was developed in the nineteenth century essentially as a corollary of a principle of mechanics.

Consider, for example, a pendulum whose mass swings back and forth between the points A and B. At these points the mass m is higher by the amount h than it is at C, the lowest point of the path (see drawing). At C, on the other hand,

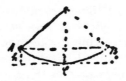

DRAWING FROM DR. EINSTEIN'S MANUSCRIPT

the lifting height has disappeared and instead of it the mass
has a velocity v. It is as though the lifting height could be
converted entirely into velocity, and vice versa. The exact
relation would be expressed as $mgh = \frac{m}{2}v^2$, with g representing the acceleration of gravity. What is interesting here is
that this relation is independent of both the length of the
pendulum and the form of the path through which the mass
moves.

The significance is that something remains constant throughout the process, and that something is energy. At A and at
B it is an energy of position, or "potential" energy; at C
it is an energy of motion, or "kinetic" energy. If this concept is correct, then the sum $mgh + m\frac{v^2}{2}$ must have the same
value for any position of the pendulum, if h is understood to
represent the height above C, and v the velocity at that point
in the pendulum's path. And such is found to be actually the
case. The generalization of this principle gives us the law of
the conservation of mechanical energy. But what happens
when friction stops the pendulum?

The answer to that was found in the study of heat phenomena. This study, based on the assumption that heat is an
indestructible substance which flows from a warmer to a colder
object, seemed to give us a principle of the "conservation of
heat." On the other hand, from time immemorial it has been
known that heat could be produced by friction, as in the fire-
making drills of the Indians. The physicists were for long
unable to account for this kind of heat "production." Their
difficulties were overcome only when it was successfully established that, for any given amount of heat produced by friction,

an exactly proportional amount of energy had to be expended. Thus did we arrive at a principle of the "equivalence of work and heat." With our pendulum, for example, mechanical energy is gradually converted by friction into heat.

In such fashion the principles of the conservation of mechanical and thermal energies were merged into one. The physicists were thereupon persuaded that the conservation principle could be further extended to take in chemical and electromagnetic processes—in short, could be applied to all fields. It appeared that in our physical system there was a sum total of energies that remained constant through all changes that might occur.

Now for the principle of the conservation of mass. Mass is defined by the resistance that a body opposes to its acceleration (inert mass). It is also measured by the weight of the body (heavy mass). That these two radically different definitions lead to the same value for the mass of a body is, in itself, an astonishing fact. According to the principle—namely, that masses remain unchanged under any physical or chemical changes—the mass appeared to be the essential (because unvarying) quality of matter. Heating, melting, vaporization, or combining into chemical compounds would not change the total mass.

Physicists accepted this principle up to a few decades ago. But it proved inadequate in the face of the special theory of relativity. It was therefore merged with the energy principle —just as, about sixty years before, the principle of the conservation of mechanical energy had been combined with the principle of the conservation of heat. We might say that the principle of the conservation of energy, having previously swallowed up that of the conservation of heat, now proceeded to swallow that of the conservation of mass—and holds the field alone.

It is customary to express the equivalence of mass and energy (though somewhat inexactly) by the formula $E = mc^2$, in which c represents the velocity of light, about 186,000 miles per second. E is the energy that is contained in a stationary body; m is its mass. The energy that belongs to the mass m is equal to this mass, multiplied by the square of the enormous speed of light—which is to say, a vast amount of energy for every unit of mass.

But if every gram of material contains this tremendous energy, why did it go so long unnoticed? The answer is simple enough: so long as none of the energy is given off externally, it cannot be observed. It is as though a man who is fabulously rich should never spend or give away a cent; no one could tell how rich he was.

Now we can reverse the relation and say that an increase of E in the amount of energy must be accompanied by an increase of $\frac{E}{c^2}$ in the mass. I can easily supply energy to the mass—for instance, if I heat it by ten degrees. So why not measure the mass increase, or weight increase, connected with this change? The trouble here is that in the mass increase the enormous factor c^2 occurs in the denominator of the fraction. In such a case the increase is too small to be measured directly; even with the most sensitive balance.

For a mass increase to be measurable, the change of energy per mass unit must be enormously large. We know of only one sphere in which such amounts of energy per mass unit are released: namely, radioactive disintegration. Schematically, the process goes like this: An atom of the mass M splits into two atoms of the mass M' and M'', which separate with tremendous kinetic energy. If we imagine these two masses as brought to rest—that is, if we take this energy of motion from them— then, considered together, they are essentially poorer in energy than was the original atom. According to the equivalence principle, the mass sum $M' + M''$ of the disintegration products must also be somewhat smaller than the original mass M of the disintegrating atom—in contradiction to the old principle of the conservation of mass. The relative difference of the two is on the order of one-tenth of one percent.

Now, we cannot actually weigh the atoms individually. However, there are indirect methods for measuring their weights exactly. We can likewise determine the kinetic energies that are transferred to the disintegration products M' and M''. Thus it has become possible to test and confirm the equivalence formula. Also, the law permits us to calculate in advance, from precisely determined atomic weights, just how much energy will be released with any atomic disintegration we have in mind. The law says nothing, of course, as to whether—or how —the disintegration reaction can be brought about.

What takes place can be illustrated with the help of our rich man. The atom M is a rich miser who, during his life, gives away no money (*energy*). But in his will he bequeaths his fortune to his sons M' and M'', on condition that they give to the community a small amount, less than one-thousandth of the whole estate (*energy or mass*). The sons together have somewhat less than the father had (*the mass sum $M' + M''$ is somewhat smaller than the mass M of the radioactive atom*). But the part given to the community, though relatively small, is still so enormously large (*considered as kinetic energy*) that it brings with it a great threat of evil. Averting that threat has become the most urgent problem of our time.

ON THE GENERALIZED THEORY OF GRAVITATION

From Scientific American, *Vol. 182, No. 4. April, 1950.*

The editors of *Scientific American* have asked me to write about my recent work which has just been published. It is a mathematical investigation concerning the foundations of field physics.

Some readers may be puzzled: didn't we learn all about the foundations of physics when we were still at school? The answer is "yes" or "no," depending on the interpretation. We have become acquainted with concepts and general relations that enable us to comprehend an immense range of experiences and make them accessible to mathematical treatment. In a certain sense these concepts and relations are probably even final. This is true, for example, of the laws of light refraction, of the relations of classical thermodynamics as far as it is based on the concepts of pressure, volume, temperature, heat, and work, and of the hypothesis of the non-existence of a perpetual motion machine.

What, then, impels us to devise theory after theory? Why do we devise theories at all? The answer to the latter question is simply: because we enjoy "comprehending," i.e., reducing phenomena by the process of logic to something already known or (apparently) evident. New theories are first of all necessary when we encounter new facts which cannot be "explained" by existing theories. But this motivation for setting up new theories is, so to speak, trivial, imposed from without. There is another, more subtle motive of no less importance. This is the striving toward unification and simplification of the premises of the theory as a whole (i.e., Mach's principle of economy, interpreted as a logical principle).

There exists a passion for comprehension, just as there exists a passion for music. That passion is rather common in children, but gets lost in most people later on. Without this passion, there would be neither mathematics nor natural science. Time and again the passion for understanding has led to the illusion that man is able to comprehend the objective world rationally, by pure thought, without any empirical foundations—in short, by metaphysics. I believe that every true theorist is a kind of tamed metaphysicist, no matter how pure a "positivist" he may fancy himself. The metaphysicist believes that the logically simple is also the real. The tamed metaphysicist believes that not all that is logically simple is embodied in experienced reality, but that the totality of all sensory experience can be "comprehended" on the basis of a conceptual system built on premises of great simplicity. The skeptic will say that this is a "miracle creed." Admittedly so, but it is a miracle creed which has been borne out to an amazing extent by the development of science.

The rise of atomism is a good example. How may Leucippus have conceived this bold idea? When water freezes and becomes ice—apparently something entirely different from water—why is it that the thawing of the ice forms something which seems indistinguishable from the original water? Leucippus is puzzled and looks for an "explanation." He is driven to the conclusion that in these transitions the "essence" of the thing has not changed at all. Maybe the thing consists of immutable particles and the change is only a change in their spatial arrangement. Could it not be that the same is true of all material objects which emerge again and again with nearly identical qualities?

This idea is not entirely lost during the long hibernation of Occidental thought. Two thousand years after Leucippus, Bernoulli wonders why gas exerts pressure on the walls of a container. Should this be "explained" by mutual repulsion of the parts of the gas, in the sense of Newtonian mechanics? This hypothesis appears absurd, for the gas pressure depends on the temperature, all other things being equal. To assume that the Newtonian forces of interaction depend on temperature is contrary to the spirit of Newtonian mechanics. Since Bernoulli is aware of the concept of atomism, he is bound to conclude that the atoms or (molecules) collide with the walls of the container and in doing so exert pressure. After all, one has to assume that atoms are in motion; how else can one account for the varying temperature of gases?

A simple mechanical consideration shows that this pressure depends only on the kinetic energy of the particles and on their density in space. This should have led the physicists of that age to the conclusion that heat consists in random motion of the atoms. Had they taken this consideration as seriously as it deserved to be taken, the development of the theory of heat—in particular the discovery of the equivalence of heat and mechanical energy—would have been considerably facilitated.

This example is meant to illustrate two things. The theoretical idea (atomism in this case) does not arise apart from and independent of experience; nor can it be derived from experience by a purely logical procedure. It is produced by a creative act. Once a theoretical idea has been acquired, one does well to hold fast to it until it leads to an untenable conclusion.

As for my latest theoretical work, I do not feel justified in giving a detailed account of it before a wide group of readers interested in science. That should be done only with theories which have been adequately confirmed by experience. So far it is primarily the simplicity of its premises and its intimate connection with what is already known (viz., the laws of the pure gravitational field) that speak in favor of the theory to be discussed here. It may, however, be of interest to a wide group of readers to become acquainted with the train of thought which can lead to endeavors of such an extremely speculative nature. Moreover, it will be shown what kinds of difficulties are encountered and in what sense they have been overcome.

In Newtonian physics the elementary theoretical concept on which the theoretical description of material bodies is based is the material point, or particle. Thus matter is considered *a priori* to be discontinuous. This makes it necessary to consider the action of material points on one another as "action at a distance." Since the latter concept seems quite contrary to everyday experience, it is only natural that the contemporaries of Newton—and indeed Newton himself—found it difficult to accept. Owing to the almost miraculous success of the Newtonian system, however, the succeeding generations of physicists became used to the idea of action at a distance. Any doubt was buried for a long time to come.

But when, in the second half of the nineteenth century, the laws of electrodynamics became known, it turned out that these laws could not be satisfactorily incorporated into the Newtonian system. It is fascinating to muse: Would Faraday have discovered the law of electromagnetic induction if he had received a regular college education? Unencumbered by the traditional way of thinking, he felt that the introduction of the "field" as an independent element of reality helped him to coordinate the experimental facts. It was Maxwell who fully comprehended the significance of the field concept; he made the fundamental discovery that the laws of electrodynamics found their natural expression in the differential equations for the electric and magnetic fields. These equations implied the existence of waves, whose properties corresponded to those of light as far as they were known at that time.

This incorporation of optics into the theory of electromagnetism represents one of the greatest triumphs in the striving toward unification of the foundations of physics; Maxwell achieved this unification by purely theoretical arguments, long before it was corroborated by Hertz's experimental work. The new insight made it possible to dispense with the hypothesis of action at a distance, at least in the realm of electromagnetic phenomena; the intermediary field now appeared as the only carrier of electromagnetic interaction between bodies, and the field's behavior was completely determined by contiguous processes, expressed by differential equations.

Now a question arose: Since the field exists even in a vacuum, should one conceive of the field as a state of a "carrier," or should it rather be endowed with an independent existence not reducible to anything else? In other words, is there an "ether" which carries the field; the ether being considered in the undulatory state, for example, when it carries light waves?

The question has a natural answer: Because one cannot dispense with the field concept, it is preferable not to introduce in addition a carrier with hypothetical properties. However, the pathfinders who first recognized the indispensability of the field concept were still too strongly imbued with the mechanistic tradition of thought to accept unhesitatingly this simple point of view. But in the course of the following decades this view imperceptibly took hold.

The introduction of the field as an elementary concept gave rise to an inconsistency of the theory as a whole. Maxwell's theory, although adequately describing the behavior of electrically charged particles in their interaction with one another, does not explain the behavior of electrical densities, i.e., it does not provide a theory of the particles themselves. They must therefore be treated as mass points on the basis of the old theory. The combination of the idea of a continuous field with that of material points discontinuous in space appears inconsistent. A consistent field theory requires continuity of all elements of the theory, not only in time but also in space, and in all points of space. Hence the material particle has no place as a fundamental concept in a field theory. Thus even apart from the fact that gravitation is not included, Maxwell's electrodynamics cannot be considered a complete theory.

Maxwell's equations for empty space remain unchanged if

the spatial coordinates and the time are subjected to a particular kind of linear transformations—the Lorentz transformations ("covariance" with respect to Lorentz transformations). Covariance also holds, of course, for a transformation which is composed of two or more such transformations; this is called the "group" property of Lorentz transformations.

Maxwell's equations imply the "Lorentz group," but the Lorentz group does not imply Maxwell's equations. The Lorentz group may indeed be defined independently of Maxwell's equations as a group of linear transformations which leave a particular value of the velocity—the velocity of light—invariant. These transformations hold for the transition from one "inertial system" to another which is in uniform motion relative to the first. The most conspicuous novel property of this transformation group is that it does away with the absolute character of the concept of simultaneity of events distant from each other in space. On this account it is to be expected that all equations of physics are covariant with respect to Lorentz transformations (*special theory of relativity*). Thus it came about that Maxwell's equations led to a heuristic principle valid far beyond the range of the applicability or even validity of the equations themselves.

Special relativity has this in common with Newtonian mechanics: The laws of both theories are supposed to hold only with respect to certain coordinate systems: those known as "inertial systems." An inertial system is a system in a state of motion such that "force-free" material points within it are not accelerated with respect to the coordinate system. However, this definition is empty if there is no independent means for recognizing the absence of forces. But such a means of recognition does not exist if gravitation is considered as a "field."

Let A be a system uniformly accelerated with respect to an "inertial system" I. Material points, not accelerated with respect to I, are accelerated with respect to A, the acceleration of all the points being equal in magnitude and direction. They behave as if a gravitational field exists with respect to A, for it is a characteristic property of the gravitational field that the acceleration is independent of the particular nature of the body.

There is no reason to exclude the possibility of interpreting this behavior as the effect of a "true" gravitational field (*principle of equivalence*). This interpretation implies that A is an "inertial system," even though it is accelerated with respect to another inertial system. (It is essential for this argument that the introduction of independent gravitational fields is considered justified even though no masses generating the field are defined. Therefore, to Newton such an argument would not have appeared convincing.) Thus the concepts of inertial system, the law of inertia and the law of motion are deprived of their concrete meaning—not only in classical mechanics but also in special relativity. Moreover, following up this train of thought, it turns out that with respect to A time cannot be measured by identical clocks; indeed, even the immediate physical significance of coordinate differences is generally lost. In view of all these difficulties, should one not try, after all, to hold cn to the concept of the inertial system, relinquishing the attempt to explain the fundamental character of the gravitational phenomena which manifest themselves in the Newtonian system as the equivalence of inert and gravitational mass? Those who trust in the comprehensibility of nature must answer: No.

This is the gist of the principle of equivalence: In order to account for the equality of inert and gravitational mass within the theory it is necessary to admit non-linear transformations of the four coordinates. That is, the group of Lorentz transformations and hence the set of the "permissible" coordinate systems has to be extended.

What group of coordinate transformations can then be substituted for the group of Lorentz transformations? Mathematics suggests an answer which is based on the fundamental investigations of Gauss and Riemann: namely, that the appropriate substitute is the group of all continuous (analytical) transformations of the coordinates. Under these transformations the only thing that remains invariant is the fact that neighboring points have nearly the same coordinates; the coordinate system expresses only the topological order of the points in space (including its four-dimensional character). The equations expressing the laws of nature must be covariant with respect to all continuous transformations of the coordinates. This is the principle of *general relativity*.

The procedure just described overcomes a deficiency in the foundations of mechanics which had already been noticed by Newton and was criticized by Leibnitz and, two centuries later, by Mach: inertia resists acceleration, but acceleration relative to what? Within the frame of classical mechanics the only answer is: inertia resists acceleration *relative to space*. This is a physical property of space—space acts on objects, but objects do not act on space. Such is probably the deeper meaning of Newton's assertion *spatium est absolutum* (space is absolute). But the idea disturbed some, in particular Leibnitz, who did not ascribe an independent existence to space but considered it merely a property of "things" (contiguity of physical objects). Had his justified doubts won out at that time, it hardly would have been a boon to physics, for the empirical and theoretical foundations necessary to follow up his idea were not available in the seventeenth century.

According to general relativity, the concept of space detached from any physical content does not exist. The physical reality of space is represented by a field whose components are continuous functions of four independent variables—the coordinates of space and time. It is just this particular kind of dependence that expresses the spatial character of physical reality.

Since the theory of general relativity implies the representation of physical reality by a *continuous* field, the concept of particles or material points cannot play a fundamental part, nor can the concept of motion. The particle can only appear as a limited region in space in which the field strength or the energy density are particularly high.

A relativistic theory has to answer two questions: (1) What is the mathematical character of the field? (2) What equations hold for this field?

Concerning the first question: from the mathematical point of view the field is essentially characterized by the way its components transform if a coordinate transformation is applied. Concerning the second question: the equations must determine the field *to a sufficient extent* while satisfying the postulates of general relativity. Whether or not this requirement can be satisfied depends on the choice of the field-type.

The attempt to comprehend the correlations among the empirical data on the basis of such a highly abstract program may at first appear almost hopeless. The procedure amounts, in fact, to putting the question: what most simple property can be required from what most simple object (field) while preserving the principle of general relativity? Viewed from the standpoint of formal logic, the dual character of the question appears calamitous, quite apart from the vagueness of the concept "simple." Moreover, from the standpoint of physics there is nothing to warrant the assumption that a theory which is "logically simple" should also be "true."

Yet every theory is speculative. When the basic concepts of a theory are comparatively "close to experience" (e.g., the concepts of force, pressure, mass), its speculative character is not so easily discernible. If, however, a theory is such as to require the application of complicated logical processes in order to reach conclusions from the premises that can be confronted with observation, everybody becomes conscious of the speculative nature of the theory. In such a case an almost irresistible feeling of aversion arises in people who are inexperienced in epistemological analysis and who are unaware of the precarious nature of theoretical thinking in those fields with which they are familiar.

On the other hand, it must be conceded that a theory has an important advantage if its basic concepts and fundamental hypotheses are "close to experience," and greater confidence in such a theory is certainly justified. There is less danger of going completely astray, particularly since it takes so much less time and effort to disprove such theories by experience. Yet more and more, as the depth of our knowledge increases, we must give up this advantage in our quest for logical simplicity and uniformity in the foundations of physical theory. It has to be admitted that general relativity has gone further than previous physical theories in relinquishing "closeness to experience" of fundamental concepts in order to attain logical simplicity. This holds already for the theory of gravitation, and it is even more true of the new generalization, which is an attempt to comprise the properties of the total field. In the generalized theory the procedure of deriving from the premises of the theory conclusions that can be confronted with empirical data is so difficult that so far no such result has been obtained. In favor of this theory are, at this point, its logical simplicity and its "rigidity." Rigidity means here that the theory is either true or false, but not modifiable.

The greatest inner difficulty impeding the development of the theory of relativity is the dual nature of the problem, indicated by the two questions we have asked. This duality is the reason why the development of the theory has taken place in two steps so widely separated in time. The first of these steps, the theory of gravitation, is based on the principle of equivalence discussed above and rests on the following consideration: According to the theory of special relativity, light has a constant velocity of propagation. If a light ray in a vacuum starts from a point, designated by the coordinates x_1, x_2 and x_3 in a three dimensional coordinate system, at the time x_4, it spreads as a spherical wave and reaches a neighboring point $(x_1 + dx_1, x_2 + dx_2, x_3 + dx_3)$ at the time $x_4 + dx_4$. Introducing the velocity of light, c, we write the expression:

$$\sqrt{dx_1{}^2 + dx_2{}^2 + dx_3{}^2} = c\,dx_4$$

This can also be written in the form:

$$dx_1{}^2 + dx_2{}^2 + dx_3{}^2 - c^2\,dx_4{}^2 = 0$$

This expression represents an objective relation between neighboring space-time points in four dimensions, and it holds for all inertial systems, provided the coordinate transformations are restricted to those of special relativity. The relation loses this form, however, if arbitrary continuous transformations of the coordinates are admitted in accordance with the principle of general relativity. The relation then assumes the more general form:

$$\sum_{ik} g_{ik}\,dx_i\,dx_k = 0$$

The g_{ik} are certain functions of the coordinates which transform in a definite way if a continuous coordinate transformation is applied. According to the principle of equivalence, these g_{ik}

functions describe a particular kind of gravitational field: a field which can be obtained by transformation of "field-free" space. The g_{ik} satisfy a particular law of transformation. Mathematically speaking, they are the components of a "tensor" with a property of symmetry which is preserved in all transformations; the symmetrical property is expressed as follows:

$$g_{ik} = g_{ki}$$

The idea suggests itself: May we not ascribe objective meaning to such a symmetrical tensor, even though the field *cannot* be obtained from the empty space of special relativity by a mere coordinate transformation? Although we cannot expect that such a symmetrical tensor will describe the most general field, it may well describe the particular case of the "pure gravitational field." Thus it is evident what kind of field, at least for a special case, general relativity has to postulate: a symmetrical tensor field.

Hence only the second question is left: What kind of general covariant field law can be postulated for a symmetrical tensor field?

This question has not been difficult to answer in our time, since the necessary mathematical conceptions were already at hand in the form of the metric theory of surfaces, created a century ago by Gauss and extended by Riemann to manifolds of an arbitrary number of dimensions. The result of this purely formal investigation has been amazing in many respects. The differential equations which can be postulated as field law for g_{ik} cannot be of lower than second order, i.e., they must at least contain the second derivatives of the g_{ik} with respect to the coordinates. Assuming that no higher than second derivatives appear in the field law, *it is mathematically determined by the principle of general relativity.* The system of equations can be written in the form:

$$R_{ik} = 0$$

The R_{ik} transform in the same manner as the g_{ik}, i.e., they too form a symmetrical tensor.

These differential equations completely replace the Newtonian theory of the motion of celestial bodies provided the masses are represented as singularities of the field. In other words, they contain the law of force as well as the law of motion while eliminating "inertial systems."

The fact that the masses appear as singularities indicates that these masses themselves cannot be explained by symmetrical g_{ik} fields, or "gravitational fields." Not even the fact that only *positive* gravitating masses exist can be deduced from this theory. Evidently a complete relativistic field theory must be based on a field of more complex nature, that is, a generalization of the symmetrical tensor field.

Before considering such a generalization, two remarks pertaining to gravitational theory are essential for the explanation to follow.

The first observation is that the principle of general relativity imposes exceedingly strong restrictions on the theoretical possibilities. Without this restrictive principle it would be practically impossible for anybody to hit on the gravitational equations, not even by using the principle of special relativity, even though one knows that the field has to be described by a symmetrical tensor. No amount of collection of facts could lead to these equations unless the principle of general relativity were used. This is the reason why all attempts to obtain a deeper knowledge of the foundations of physics seem doomed to me unless the basic concepts are in accordance with general relativity from the beginning. This situation makes it difficult to use our empirical knowledge, however comprehensive, in looking for the fundamental concepts and relations of physics, and it forces us to apply free speculation to a much greater extent than is presently assumed by most physicists. I do not see any reason to assume that the heuristic significance of the principle of general relativity is restricted to gravitation and that the rest of physics can be dealt with separately on the basis of special relativity, with the hope that later on the whole may be fitted consistently into a general relativistic scheme. I do not think that such an attitude, although historically understandable, can be objectively justified. The comparative smallness of what we know today as gravitational effects is not a conclusive reason for ignoring the principle of general relativity in theoretical investigations of a fundamental character. In other words, I do not believe that it is justifiable to ask: what would physics look like without gravitation?

The second point we must note is that the equations of gravitation are ten differential equations for the ten components of the symmetrical tensor g_{ik}. In the case of a non-general relativistic theory, a system is ordinarily not overdetermined if the number of equations is equal to the number of unknown functions. The manifold of solutions is such that within the general solution a certain number of functions of three variables can be chosen arbitrarily. For a general relativistic theory this cannot be expected as a matter of course. Free choice with respect to the coordinate system implies that out of the ten functions of a solution, or components of the field, four can be made to assume prescribed values by a suitable choice of the coordinate system. In other words, the principle of general relativity implies that the number of functions to be determined by differential equations is not 10 but $10 - 4 = 6$. For these six functions only six independent differential equations may be postulated. Only six out of the ten differential equations of the gravitational field ought to be independent of each other, while the remaining four must be connected to those six by means of four relations (identities). And indeed there exist among the left-hand sides, R_{ik}, of the ten gravitational equations four identities— "Bianchi's identities"—which assure their "compatibility."

In a case like this—when the number of field variables is equal to the number of differential equations—compatibility is always assured if the equations can be obtained from a variational principle. This is indeed the case for the gravitational equations.

However, the ten differential equations cannot be entirely replaced by six. The system of equations is indeed "overdetermined," but due to the existence of the identities it is overdetermined in such a way that its compatibility is not lost, i.e., the manifold of solutions is not critically restricted. The fact that the equations of gravitation imply the law of motion for the masses is intimately connected with this (permissible) overdetermination.

After this preparation it is now easy to understand the nature of the present investigation without entering into the details of its mathematics. The problem is to set up a relativistic theory for the total field. The most important clue to its solution is that there exists already the solution for the special case of the pure gravitational field. The theory we are looking for must therefore be a generalization of the theory of the gravitational field. The first question is: what is the natural generalization of the symmetrical tensor field?

This question cannot be answered by itself, but only in connection with the other question: what generalization of the field is going to provide the most natural theoretical system? The answer on which the theory under discussion is based is that the symmetrical tensor field must be replaced by a non-symmetrical one. This means that the condition $g_{ik} = g_{ki}$ for the field components must be dropped. In that case the field has sixteen instead of ten independent components.

There remains the task of setting up the relativistic differential equations for a non-symmetrical tensor field. In the attempt to solve this problem one meets with a difficulty which does not arise in the case of the symmetrical field. The principle of general relativity does not suffice to determine completely the field equations, mainly because the transformation law of the symmetrical part of the field alone does not involve the components of the antisymmetrical part or *vice versa*. Probably this is the reason why this kind of generalization of the field has hardly ever been tried before. The combination of the two parts of the field can only be shown to be a natural procedure if in the formalism of the theory only the total field plays a role, and not the symmetrical and antisymmetrical parts separately.

It turned out that this requirement can indeed be satisfied in a natural way. But even this requirement, together with the principle of general relativity, is still not sufficient to determine uniquely the field equations. Let us remember that the system of equations must satisfy a further condition: the equations must be compatible. It has been mentioned above that this condition is satisfied if the equations can be derived from a variational principle.

This has indeed been achieved, although not in so natural

a way as in the case of the symmetrical field. It has been disturbing to find that it can be achieved in two different ways. These variational principles furnished two systems of equations —let us denote them by E_1 and E_2—which were different from each other (although only slightly so), each of them exhibiting specific imperfections. Consequently even the condition of compatibility was insufficient to determine the system of equations uniquely.

It was, in fact, the formal defects of the systems E_1 and E_2 that indicated a possible way out. There exists a third system of equations, E_3, which is free of the formal defects of the systems E_1 and E_2 and represents a combination of them in the sense that every solution of E_3 is a solution of E_1 as well as of E_2. This suggests that E_3 may be the system we have been looking for. Why not postulate E_3, then, as the system of equations? Such a procedure is not justified without further analysis, since the compatibility of E_1 and that of E_2 do not imply compatibility of the stronger system E_3, where the number of equations exceeds the number of field components by four.

An independent consideration shows that irrespective of the question of compatibility the stronger system, E_3, is the only really natural generalization of the equations of gravitation.

But E_3 is not a compatible system in the same sense as are the systems E_1 and E_2, whose compatibility is assured by a sufficient number of identities, which means that every field that satisfies the equations for a definite value of the time has a continuous extension representing a solution in four-dimensional space. The system E_3, however, is not extensible in the same way. Using the language of classical mechanics, we might say: in the case of the system E_3 the "initial condition" cannot be freely chosen. What really matters is the answer to the question: is the manifold of solutions for the system E_3 as extensive as must be required for a physical theory? This purely mathematical problem is as yet unsolved.

The skeptic will say: "It may well be true that this system of equations is reasonable from a logical standpoint. But this does not prove that it corresponds to nature." You are right, dear skeptic. Experience alone can decide on truth. Yet we have achieved something if we have succeeded in formulating a meaningful and precise question. Affirmation or refutation will not be easy, in spite of an abundance of known empirical facts. The derivation, from the equations, of conclusions which can be confronted with experience will require painstaking efforts and probably new mathematical methods.

MESSAGE TO THE ITALIAN SOCIETY FOR THE ADVANCEMENT OF SCIENCE

Sent to the forty-second meeting of the "Società Italiana per il Progresse de la Scienze," Lucca (Italy), 1950. Published in English in the Unesco periodical, Impact, *Autumn, 1950.*

Let me first thank you most sincerely for your kindness in inviting me to attend the meeting of the "Society for the Advancement of Science." I should gladly have accepted the invitation if my health had permitted me to do so. All I can do under the circumstances is to address you briefly from my home across the ocean. In doing so, I am under no illusion that I have something to say which would actually enlarge your insight and understanding. However, we are living in a period of such great external and internal insecurity and with such a lack of firm objectives that the mere confession of our convictions may be of significance even if these convictions, like all value judgments, cannot be proven through logical deductions.

There arises at once the question: should we consider the search for truth or, more modestly expressed, our efforts to understand the knowable universe through constructive logical thought as an autonomous objective of our work? Or should our search for truth be subordinated to some other objective, for example, to a "practical" one? This question cannot be decided on a logical basis. The decision, however, will have considerable influence upon our thinking and our moral judgment, provided that it is born out of deep and unshakable conviction. Let me then make a confession: for myself, the struggle to gain more insight and understanding is one of those independent objectives without which a thinking individual would find it impossible to have a conscious, positive attitude toward life. It is the very essence of our striving for understanding that, on the one hand, it attempts to encompass the great and complex variety of man's experience, and that on the other, it looks for simplicity and economy in the basic assumptions. The belief that these two objectives can exist side by side is, in view of the primitive state of our scientific knowledge, a matter of faith. Without such faith I could not have a strong and unshakable conviction about the independent value of knowledge.

This, in a sense, religious attitude of a man engaged in scientific work has some influence upon his whole personality. For apart from the knowledge which is offered by accumulated experience and from the rules of logical thinking, there exists in principle for the man in science no authority whose decisions and statements could have in themselves a claim to "Truth." This leads to the paradoxical situation that a person who devotes all his strength to objective matters will develop, from a social point of view, into an extreme individualist who, at least in principle, has faith in nothing but his own judgment. It is quite possible to assert that intellectual individualism and scientific eras emerged simultaneously in history and have remained inseparable ever since.

Someone may suggest that the man of science as sketched in these sentences is no more than an abstraction which actually does not exist in this world, not unlike the *homo oeconomicus* of classical economics. However, it seems to me that science as we know it today could not have emerged and could not have remained alive if many individuals, during many centuries, would not have come very close to the ideal.

Of course, not everybody who has learned to use tools and methods which, directly or indirectly, appear to be "scientific" is to me a man of science. I refer only to those individuals in whom scientific mentality is truly alive.

What, then, is the position of today's man of science as a member of society? He obviously is rather proud of the fact that the work of scientists has helped to change radically the economic life of men by almost completely eliminating muscular work. He is distressed by the fact that the results of his

scientific work have created a threat to mankind since they have
fallen into the hands of morally blind exponents of political
power. He is conscious of the fact that technological methods
made possible by his work have led to a concentration of economic and also of political power in the hands of small minorities
which have come to dominate completely the lives of the masses
of people who appear more and more amorphous. But even
worse: the concentration of economic and political power in
few hands has not only made the man of science dependent
economically; it also threatens his independence from within;
the shrewd methods of intellectual and psychic influences which
it brings to bear will prevent the development of really independent personalities.

Thus the man of science, as we can observe with our own eyes,
suffers a truly tragic fate. Striving in great sincerity for clarity
and inner independence, he himself, through his sheer superhuman efforts, has fashioned the tools which are being used to
make him a slave and to destroy him also from within. He
cannot escape being muzzled by those who have the political
power in their hands. As a soldier he is forced to sacrifice his
own life and to destroy the lives of others even when he is convinced of the absurdity of such sacrifices. He is fully aware of
the fact that universal destruction is unavoidable since the historical development has led to the concentration of all economic,
political, and military power in the hands of national states. He
also realizes that mankind can be saved only if a supranational
system, based on law, would be created to eliminate for good the
methods of brute force. However, the man of science has slipped
so much that he accepts the slavery inflicted upon him by
national states as his inevitable fate. He even degrades himself
to such an extent that he helps obediently in the perfection of
the means for the general destruction of mankind.

Is there really no escape for the man of science? Must he
really tolerate and suffer all these indignities? Is the time gone
forever when, aroused by his inner freedom and the independence of his thinking and his work, he had a chance of enlighten-
ing and enriching the lives of his fellow human beings? In
placing his work too much on an intellectual basis, has he not

forgotten about his responsibility and dignity? My answer is: while it is true that an inherently free and scrupulous person may be destroyed, such an individual can never be enslaved or used as a blind tool.

If the man of science of our own days could find the time and the courage to think over honestly and critically his situation and the tasks before him and if he would act accordingly, the possibilities for a sensible and satisfactory solution of the present dangerous international situation would be considerably improved.

MESSAGE ON THE 410TH ANNIVERSARY OF THE DEATH OF COPERNICUS

On the occasion of the commemoration evening held at Columbia University, New York, in December, 1953.

We are honoring today, with joy and gratitude, the memory of a man who, more than almost anyone else, contributed to the liberation of the mind from the chains of clerical and scientific dominance in the Occident.

It is true that some scholars in the classic Greek period had become convinced that the earth is not the natural center of the world. But this comprehension of the universe could not gain real recognition in antiquity. Aristotle and the Greek school of astronomers continued to adhere to the geocentric conception, and hardly anyone had any doubt about it.

A rare independence of thought and intuition as well as a mastery of the astronomical facts, not easily accessible in those days, were necessary to expound the superiority of the heliocentric conception convincingly. This great accomplishment of Copernicus not only paved the way to modern astronomy; it also helped to bring about a decisive change in man's attitude toward the cosmos. Once it was recognized that the earth was not the center of the world, but only one of the smaller planets, the illusion of the central significance of man himself became untenable. Hence, Copernicus, through his work and the greatness of his personality, taught man to be modest.

No nation should find pride in the fact that such a man

developed in its midst. For national pride is quite a petty weakness which is hardly justifiable in face of a man of such inner independence as Copernicus.

RELATIVITY AND THE PROBLEM OF SPACE

From the revised edition of Relativity, the Special and the General Theory: A Popular Exposition. *Translated by Robert W. Lawson. London: Methuen, 1954.*

It is characteristic of Newtonian physics that it has to ascribe independent and real existence to space and time as well as to matter, for in Newton's law of motion the concept of acceleration appears. But in this theory, acceleration can only denote "acceleration with respect to space." Newton's space must thus be thought of as "at rest," or at least as "unaccelerated," in order that one can consider the acceleration, which appears in the law of motion, as being a magnitude with any meaning. Much the same holds with time, which of course likewise enters into the concept of acceleration. Newton himself and his most critical contemporaries felt it to be disturbing that one had to ascribe physical reality both to space itself as well as to its state of motion; but there was at that time no other alternative, if one wished to ascribe to mechanics a clear meaning.

It is indeed an exacting requirement to have at all to ascribe physical reality to space, and especially to empty space. Time and again since remotest times philosophers have resisted such a presumption. Descartes argued somewhat on these lines: space is identical with extension, but extension is connected with bodies; thus there is no space without bodies and hence no empty space. The weakness of this argument lies primarily in what follows. It is certainly true that the concept of extension owes its origin to our experiences of laying out or bringing into contact solid bodies. But from this it cannot be concluded that the concept of extension may not be justified in cases which have not themselves given rise to the formation of this concept. Such an enlargement of concepts can be justified indirectly by its value for the comprehension of empirical results. The assertion that extension is confined to bodies is therefore of itself certainly unfounded. We shall see later, however, that the general theory of relativity confirms Descartes' conception in a roundabout way. What brought Descartes to his seemingly odd view was certainly the feeling that, without compelling necessity, one ought not to ascribe reality to a thing like space, which is not capable of being "directly experienced." *

The psychological origin of the idea of space, or of the necessity for it, is far from being so obvious as it may appear to be on the basis of our customary habit of thought. The old geometers deal with conceptual objects (straight line, point, surface), but not really with space as such, as was done later in analytical geometry. The idea of space, however, is suggested by certain primitive experiences. Suppose that a box has been constructed. Objects can be arranged in a certain way inside the box, so that it becomes full. The possibility of such arrangements is a property of the material object "box," something that is given with the box, the "space enclosed" by the box. This is something which is different for different boxes, something that is thought quite naturally as being independent of whether or not, at any moment, there are any objects at all in the box. When there are no objects in the box, its space appears to be "empty."

So far, our concept of space has been associated with the box. It turns out, however, that the storage possibilities that make up the box-space are independent of the thickness of the walls of the box. Cannot this thickness be reduced to zero, without the "space" being lost as a result? The naturalness of such a limiting process is obvious, and now there remains for our thought the space without the box, a self-evident thing, yet it appears to be so unreal if we forget the origin of this concept. One can understand that it was repugnant to Descartes to consider space as independent of material objects, a thing that might exist without matter.† (At the same time, this does not prevent him from treating space as a fundamental concept in his analytical geometry.) The drawing of attention to the vacuum in a mer-

---

* This expression is to be taken *cum grano salis*.

† Kant's attempt to remove the embarrassment by denial of the objectivity of space can, however, hardly be taken seriously. The possibilities of packing inherent in the inside space of a box are objective in the same sense as the box itself, and as the objects which can be packed inside it.

cury barometer has certainly disarmed the last of the Cartesians. But it is not to be denied that, even at this primitive stage, something unsatisfactory clings to the concept of space, or to space thought of as an independent real thing.

The ways in which bodies can be packed into space (box) are the subject of three-dimensional Euclidean geometry, whose axiomatic structure readily deceives us into forgetting that it refers to realizable situations.

If now the concept of space is formed in the manner outlined above, and following on from experience about the "filling" of the box, then this space is primarily a *bounded* space. This limitation does not appear to be essential, however, for apparently a larger box can always be introduced to enclose the smaller one. In this way space appears as something unbounded.

I shall not consider here how the concepts of the three-dimensional and the Euclidean nature of space can be traced back to relatively primitive experiences. Rather, I shall consider first of all from other points of view the rôle of the concept of space in the development of physical thought.

When a smaller box s is situated, relatively at rest, inside the hollow space of a larger box S, then the hollow space of s is a part of the hollow space of S, and the same "space," which contains both of them, belongs to each of the boxes. When s is in motion with respect to S, however, the concept is less simple. One is then inclined to think that s encloses always the same space, but a variable part of the space S. It then becomes necessary to apportion to each box its particular space, not thought of as bounded, and to assume that these two spaces are in motion with respect to each other.

Before one has become aware of this complication, space appears as an unbounded medium or container in which material objects swim around. But it must now be remembered that there is an infinite number of spaces, which are in motion with respect to each other. The concept of space as something existing objectively and independent of things belongs to pre-scientific thought, but not so the idea of the existence of an infinite number of spaces in motion relatively to each other. This latter idea is indeed logically unavoidable, but is far from having played a considerable rôle even in scientific thought.

But what about the psychological origin of the concept of time? This concept is undoubtedly associated with the fact of "calling to mind," as well as with the differentiation between sense experiences and the recollection of these. Of itself it is doubtful whether the differentiation between sense experience and recollection (or a mere mental image) is something psychologically directly given to us. Everyone has experienced that he has been in doubt whether he has actually experienced something with his senses or has simply dreamed about it. Probably the ability to discriminate between these alternatives first comes about as the result of an activity of the mind creating order.

An experience is associated with a "recollection," and it is considered as being "earlier" in comparison with "present experiences." This is a conceptual ordering principle for recollected experiences, and the possibility of its accomplishment gives rise to the subjective concept of time, i.e., that concept of time which refers to the arrangement of the experiences of the individual.

What do we mean by rendering objective the concept of time? Let us consider an example. A person A ("I") has the experience "it is lightning." At the same time the person A also experiences such a behavior of the person B as brings the behavior of B into relation with his own experience "it is lightning." Thus it comes about that A associates with B the experience "it is lightning." For the person A the idea arises that other persons also participate in the experience "it is lightning." "It is lightning" is now no longer interpreted as an exclusively personal experience, but as an experience of other persons (or eventually only as a "potential experience"). In this way arises the interpretation that "it is lightning," which originally entered into the consciousness as an "experience," is now also interpreted as an (objective) "event." It is just the sum total of all events that we mean when we speak of the "real external world."

We have seen that we feel ourselves impelled to ascribe a temporal arrangement to our experiences, somewhat as follows. If β is later than α and γ later than β, then γ is also later than α

("sequence of experiences"). Now what is the position in this respect with the "events" which we have associated with the experiences? At first sight it seems obvious to assume that a temporal arrangement of events exists which agrees with the temporal arrangement of the experiences. In general, and unconsciously this was done, until skeptical doubts made themselves felt.* In order to arrive at the idea of an objective world, an additional constructive concept still is necessary: the event is localized not only in time, but also in space.

In the previous paragraphs we have attempted to describe how the concepts space, time, and event can be put psychologically into relation with experiences. Considered logically, they are free creations of the human intelligence, tools of thought, which are to serve the purpose of bringing experiences into relation with each other, so that in this way they can be better surveyed. The attempt to become conscious of the empirical sources of these fundamental concepts should show to what extent we are actually bound to these concepts. In this way we become aware of our freedom, of which, in case of necessity, it is always a difficult matter to make sensible use.

We still have something essential to add to this sketch concerning the psychological origin of the concepts space-time-event (we will call them more briefly "space-like," in contrast to concepts from the psychological sphere). We have linked up the concept of space with experiences using boxes and the arrangement of material objects in them. Thus this formation of concepts already presupposes the concept of material objects (e.g., "boxes"). In the same way persons, who had to be introduced for the formation of an objective concept of time, also play the rôle of material objects in this connection. It appears to me, therefore, that the formation of the concept of the material object must precede our concepts of time and space.

All these space-like concepts already belong to pre-scientific thought, along with concepts like pain, goal, purpose, etc., from the field of psychology. Now it is characteristic of thought in

---

* For example, the order of experiences in time obtained by acoustical means can differ from the temporal order gained visually, so that one cannot simply identify the time sequence of events with the time sequence of experiences.

physics, as of thought in natural science generally, that it endeavors in principle to make do with "space-like" concepts *alone,* and strives to express with their aid all relations having the form of laws. The physicist seeks to reduce colors and tones to vibrations; the physiologist, thought and pain to nerve processes, in such a way that the psychical element as such is eliminated from the causal nexus of existence, and thus nowhere occurs as an independent link in the causal associations. It is no doubt this attitude, which considers the comprehension of all relations by the exclusive use of only "space-like" concepts as being possible in principle, that is at the present time understood by the term "materialism" (since "matter" has lost its rôle as a fundamental concept).

Why is it necessary to drag down from the Olympian fields of Plato the fundamental ideas of thought in natural science, and to attempt to reveal their earthly lineage? Answer: In order to free these ideas from the taboo attached to them, and thus to achieve greater freedom in the formation of ideas or concepts. It is to the immortal credit of D. Hume and E. Mach that they, above all others, introduced this critical conception.

Science has taken over from pre-scientific thought the concepts space, time, and material object (with the important special case "solid body"), and has modified them and rendered them more precise. Its first significant accomplishment was the development of Euclidean geometry, whose axiomatic formulation must not be allowed to blind us to its empirical origin (the possibilities of laying out or juxtaposing solid bodies). In particular, the three-dimensional nature of space as well as its Euclidean character are of empirical origin (it can be wholly filled by like constituted "cubes").

The subtlety of the concept of space was enhanced by the discovery that there exist no completely rigid bodies. All bodies are elastically deformable and alter in volume with change in temperature. The structures, whose possible configurations are to be described by Euclidean geometry, cannot therefore be characterized without reference to the content of physics. But since physics after all must make use of geometry in the establishment of its concepts, the empirical content of geometry can be stated and tested only in the framework of the whole of physics.

In this connection atomistics must also be borne in mind, and its conception of finite divisibility; for spaces of sub-atomic extension cannot be measured up. Atomistics also compels us to give up, in principle, the idea of sharply and statically defined bounding surfaces of solid bodies. Strictly speaking, there are no *precise* laws, even in the macro-region, for the possible configurations of solid bodies touching each other.

In spite of this, no one thought of giving up the concept of space, for it appeared indispensable in the eminently satisfactory whole system of natural science. Mach, in the nineteenth century, was the only one who thought seriously of an elimination of the concept of space, in that he sought to replace it by the notion of the totality of the instantaneous distances between all material points. (He made this attempt in order to arrive at a satisfactory understanding of inertia.)

THE FIELD

In Newtonian mechanics, space and time play a dual rôle. First, they play the part of carrier or frame for things that happen in physics, in reference to which events are described by the space coordinates and the time. In principle, matter is thought of as consisting of "material points," the motions of which constitute physical happening. When matter is thought of as being continuous, this is done, as it were, provisionally in those cases where one does not wish to or cannot describe the discrete structure. In this case small parts (elements of volume) of the matter are treated similarly to material points, at least in so far as we are concerned merely with motions and not with occurrences which, at the moment, it is not possible or serves no useful purpose to attribute to motions (e.g., temperature changes, chemical processes). The second rôle of space and time was that of being an "inertial system." Inertial systems were considered to be distinguished among all conceivable systems of reference in that, with respect to them, the law of inertia claimed validity.

In this, the essential thing is that "physical reality," thought of as being independent of the subjects experiencing it, was conceived as consisting, at least in principle, of space and time on one hand, and of permanently existing material points, moving with respect to space and time, on the other. The idea of the independent existence of space and time can be expressed drastically in this way: if matter were to disappear, space and time alone would remain behind (as a kind of stage for physical happening).

This standpoint was overcome in the course of a development which, in the first place, appeared to have nothing to do with the problem of space-time, namely, the appearance of the *concept of field* and its final claim to replace, in principle, the idea of a particle (material point). In the framework of classical physics, the concept of field appeared as an auxiliary concept, in cases in which matter was treated as a continuum. For example, in the consideration of the heat conduction in a solid body, the state of the body is described by giving the temperature at every point of the body for every definite time. Mathematically, this means that the temperature T is represented as a mathematical expression (function) of the space coordinates and the time t (temperature field). The law of heat conduction is represented as a local relation (differential equation), which embraces all special cases of the conduction of heat. The temperature is here a simple example of the concept of field. This is a quantity (or a complex of quantities), which is a function of the coordinates and the time. Another example is the description of the motion of a liquid. At every point there exists at any time a velocity, which is quantitatively described by its three "components" with respect to the axes of a coordinate system (vector). The components of the velocity at a point (field components), here also are functions of the coordinates (x, y, z) and the time (t).

It is characteristic of the fields mentioned that they occur only within a ponderable mass; they serve only to describe a state of this matter. In accordance with the historical development of the field concept, where no matter was available there could also exist no field. But in the first quarter of the nineteenth century it was shown that the phenomena of the interference and the diffraction of light could be explained with astonishing accuracy when light was regarded as a wave-field, completely analogous to the mechanical vibration field in an elastic solid body. It was thus felt necessary to introduce a field, that could also exist in "empty space" in the absence of ponderable matter.

This state of affairs created a paradoxical situation, because, in accordance with its origin, the field concept appeared to be restricted to the description of states in the inside of a ponderable body. This seemed to be all the more certain, inasmuch as the conviction was held that every field is to be regarded as a state capable of mechanical interpretation, and this presupposed the presence of matter. One thus felt compelled, even in the space which had hitherto been regarded as empty, to assume everywhere the existence of a form of matter, which was called "ether."

The emancipation of the field concept from the assumption of its association with a mechanical carrier finds a place among the psychologically most interesting events in the development of physical thought. During the second half of the nineteenth century, in connection with the researches of Faraday and Maxwell, it became more and more clear that the description of electromagnetic processes in terms of field was vastly superior to a treatment on the basis of the mechanical concepts of material points. By the introduction of the field concept in electrodynamics, Maxwell succeeded in predicting the existence of electromagnetic waves, the essential identity of which with light waves could not be doubted, if only because of the equality of their velocity of propagation. As a result of this, optics was, in principle, absorbed by electrodynamics. *One* psychological effect of this immense success was that the field concept gradually won greater independence from the mechanistic framework of classical physics.

Nevertheless, it was at first taken for granted that electromagnetic fields had to be interpreted as states of the ether, and it was zealously sought to explain these states as mechanical ones. But as these efforts always met with frustration, science gradually became accustomed to the idea of renouncing such a mechanical interpretation. Nevertheless, the conviction still remained that electromagnetic fields must be states of the ether, and this was the position at the turn of the century.

The ether-theory brought with it the question: how does the ether behave from the mechanical point of view with respect to ponderable bodies? Does it take part in the motions of the bodies, or do its parts remain at rest relatively to each other? Many ingenious experiments were undertaken to decide this question. The following important facts should be mentioned in this connection: the "aberration" of the fixed stars in consequence of the annual motion of the earth, and the "Doppler effect," i.e., the influence of the relative motion of the fixed stars on the frequency of the light reaching us from them, for known frequencies of emission. The results of all these facts and experiments, except for one, the Michelson-Morley experiment, were explained by H. A. Lorentz on the assumption that the ether does not take part in the motions of ponderable bodies, and that the parts of the ether have no relative motions at all with respect to each other. Thus the ether appeared, as it were, as the embodiment of a space absolutely at rest. But the investigation of Lorentz accomplished still more. It explained all the electromagnetic and optical processes within ponderable bodies known at that time, on the assumption that the influence of ponderable matter on the electric field—and conversely—is due solely to the fact that the constituent particles of matter carry electrical charges, which share the motion of the particles. Concerning the experiment of Michelson and Morley, H. A. Lorentz showed that the result obtained at least does not contradict the theory of an ether at rest.

In spite of all these beautiful successes the state of the theory was not yet wholly satisfactory, and for the following reasons. Classical mechanics, of which it could not be doubted that it holds with a close degree of approximation, teaches the equivalence of all inertial systems or inertial "spaces" for the formulation of natural laws, i.e., the invariance of natural laws with respect to the transition from one inertial system to another. Electromagnetic and optical *experiments* taught the same thing with considerable accuracy. But the foundation of electromagnetic *theory* taught that a particular inertial system must be given preference, namely, that of the luminiferous ether at rest. This view of the theoretical foundation was much too unsatisfactory. Was there no modification that, like classical mechanics, would uphold the equivalence of inertial systems (special principle of relativity)?

The answer to this question is the special theory of relativity. This takes over from the theory of Maxwell-Lorentz the assumption of the constancy of the velocity of light in empty space. In order to bring this into harmony with the equivalence of inertial systems (special principle of relativity), the idea of the absolute character of simultaneity must be given up; in addition, the Lorentz transformations for the time and the space coordinates follow for the transition from one inertial system to another. The whole content of the special theory of relativity is included in the postulate: the laws of nature are invariant with respect to the Lorentz transformations. The importance of this requirement lies in the fact that it limits the possible natural laws in a definite manner.

What is the position of the special theory of relativity in regard to the problem of space? In the first place we must guard against the opinion that the four-dimensionality of reality has been newly introduced for the first time by this theory. Even in classical physics the event is localized by four numbers, three spatial coordinates and a time coordinate; the totality of physical "events" is thus thought of as being embedded in a four-dimensional continuous manifold. But on the basis of classical mechanics this four-dimensional continuum breaks up objectively into the one-dimensional time and into three-dimensional spatial sections, the latter of which contain only simultaneous events. This resolution is the same for all inertial systems. The simultaneity of two definite events with reference to one inertial system involves the simultaneity of these events in reference to all inertial systems. This is what is meant when we say that the time of classical mechanics is absolute. According to the special theory of relativity it is otherwise. The sum total of events which are simultaneous with a selected event exist, it is true, in relation to a particular inertial system, but no longer independently of the choice of the inertial system. The four-dimensional continuum is now no longer resolvable objectively into sections, which contain all simultaneous events; "now" loses for the spatially extended world its objective meaning. It is because of this that space and time must be regarded as a four-dimensional continuum that is objectively unresolvable, if it is desired to express the purport of objective relations without unnecessary conventional arbitrariness.

Since the special theory of relativity revealed the physical equivalence of all inertial systems, it proved the untenability of the hypothesis of an ether at rest. It was therefore necessary to renounce the idea that the electromagnetic field is to be regarded as a state of a material carrier. The field thus becomes an irreducible element of physical description, irreducible in the same sense as the concept of matter in the theory of Newton.

Up to now we have directed our attention to finding in what respect the concepts of space and time were *modified* by the special theory of relativity. Let us now focus our attention on those elements which this theory has taken over from classical mechanics. Here also, natural laws claim validity only when an inertial system is taken as the basis of space-time description. The principle of inertia and the principle of the constancy of the velocity of light are valid only with respect to an *inertial system*. The field-laws also can claim to have meaning and validity only in regard to inertial systems. Thus, as in classical mechanics, space is here also an independent component in the representation of physical reality. If we imagine matter and field to be removed, inertial space or, more accurately, this space together with the associated time remains behind. The four-dimensional structure (Minkowski-space) is thought of as being the carrier of matter and of the field. Inertial spaces, with their associated times, are only privileged four-dimensional coordinate systems that are linked together by the linear Lorentz transformations. Since there exist in this four-dimensional structure no longer any sections which represent "now" objectively, the concepts of happening and becoming are indeed not completely suspended, but yet complicated. It appears therefore more natural to think of physical reality as a four-dimensional existence, instead of, as hitherto, the *evolution* of a three-dimensional existence.

This rigid four-dimensional space of the special theory of relativity is to some extent a four-dimensional analogue of H. A. Lorentz's rigid three-dimensional ether. For this theory also the following statement is valid: the description of physical states postulates space as being initially given and as existing independently. Thus even this theory does not dispel Descartes' uneasiness concerning the independent, or indeed, the *a priori* existence of "empty space." The real aim of the elementary discussion given here is to show to what extent these doubts are overcome by the general theory of relativity.

The Concept of Space in the General Theory of Relativity

This theory arose primarily from the endeavor to understand the equality of inertial and gravitational mass. We start out from an inertial system S_1, whose space is, from the physical point of view, empty. In other words, there exists in the part of space contemplated neither matter (in the usual sense) nor a field (in the sense of the special theory of relativity). With reference to S_1 let there be a second system of reference S_2 in uniform acceleration. Then S_2 is thus not an inertial system. With respect to S_2 every test mass would move with an acceleration, which is independent of its physical and chemical nature. Relative to S_2, therefore, there exists a state which, at least to a first approximation, cannot be distinguished from a gravitational field. The following concept is thus compatible with the observable facts: S_2 is also equivalent to an "inertial system"; but with respect to S_2 a (homogeneous) gravitational field is present (about the origin of which one does not worry in this connection). Thus when the gravitational field is included in the framework of the consideration, the inertial system loses its objective significance, assuming that this "principle of equivalence" can be extended to any relative motion whatsoever of the systems of reference. If it is possible to base a consistent theory on these fundamental ideas, it will satisfy of itself the fact of the equality of inertial and gravitational mass, which is strongly confirmed empirically.

Considered four-dimensionally, a non-linear transformation of the four coordinates corresponds to the transition from S_1 to S_2. The question now arises: what kind of non-linear transformations are to be permitted, or, how is the Lorentz transformation to be generalized? In order to answer this question, the following consideration is decisive.

We ascribe to the inertial system of the earlier theory this property: differences in coordinates are measured by stationary "rigid" measuring rods, and differences in time by clocks at rest. The first assumption is supplemented by another, namely, that for the relative laying out and fitting together of measuring rods at rest, the theorems on "lengths" in Euclidean geometry hold. From the results of the special theory of relativity it is then concluded, by elementary considerations, that this direct physical interpretation of the coordinates is lost for systems of reference (S_2) accelerated relatively to inertial systems (S_1). But if this is the case, the coordinates now express only the order or rank of the "contiguity" and hence also the number of dimensions of the space, but do not express any of its metrical properties. We are thus led to extend the transformations to arbitrary continuous transformations.* This implies the general principle of relativity: Natural laws must be covariant with respect to arbitrary continuous transformations of the coordinates. This requirement (combined with that of the greatest possible logical simplicity of the laws) limits the natural laws concerned incomparably more strongly than the special principle of relativity.

This train of ideas is based essentially on the field as an independent concept. For the conditions prevailing with respect to S_2 are interpreted as a gravitational field, without the question of the existence of masses which produce this field being raised. By virtue of this train of ideas it can also be grasped why the laws of the pure gravitational field are more directly linked with the idea of general relativity than the laws

---

* This inexact mode of expression will perhaps suffice here.

for fields of a general kind (when, for instance, an electro-magnetic field is present). We have, namely, good ground for the assumption that the "field-free" Minkowski-space represents a special case possible in natural law, in fact, the simplest conceivable special case. With respect to its metrical character, such a space is characterized by the fact that $dx_1{}^2 + dx_2{}^2 + dx_3{}^2$ is the square of the spatial separation, measured with a unit gauge, of two infinitesimally neighboring points of a three-dimensional "space-like" cross section (Pythagorean theorem), whereas dx_4 is the temporal separation, measured with a suitable time gauge, of two events with common (x_1, x_2, x_3). All this simply means that an objective metrical significance is attached to the quantity

$$ds^2 = dx_1{}^2 + dx_2{}^2 + dx_3{}^2 - dx_4{}^2 \qquad (1)$$

as is readily shown with the aid of the Lorentz transformations. Mathematically, this fact corresponds to the condition that ds^2 is invariant with respect to Lorentz transformations.

If now, in the sense of the general principle of relativity, this space (cf. eq. (1)) is subjected to an arbitrary continuous transformation of the coordinates, then the objectively significant quantity ds is expressed in the new system of coordinates by the relation

$$ds^2 = g_{ik}dx_idx_k \qquad \cdot \qquad \cdot \qquad \cdot \qquad (1a)$$

which has to be summed up over the indices i and k for all combinations 11, 12, . . . up to 44. The terms g_{ik} now are not constants, but functions of the coordinates, which are determined by the arbitrarily chosen transformation. Nevertheless, the terms g_{ik} are not arbitrary functions of the new coordinates, but just functions of such a kind that the form (1a) can be transformed back again into the form (1) by a continuous transformation of the four coordinates. In order that this may be possible, the functions g_{ik} must satisfy certain general covariant equations of condition, which were derived by B. Riemann more than half a century before the formulation of the general theory of relativity ("Riemann condition"). According to the principle of equivalence, (1a) describes in general covariant form a gravitational field of a special kind, when the functions g_{ik} satisfy the Riemann condition.

It follows that the law for the pure gravitational field of a general kind must be satisfied when the Riemann condition is satisfied; but it must be weaker or less restricting than the Riemann condition. In this way the field law of pure gravitation is practically completely determined, a result which will not be justified in greater detail here.

We are now in a position to see how far the transition to the general theory of relativity modifies the concept of space. In accordance with classical mechanics and according to the special theory of relativity, space (space-time) has an existence independent of matter or field. In order to be able to describe at all that which fills up space and is dependent on the coordinates, space-time or the inertial system with its metrical properties must be thought of as existing to start with, for otherwise the description of "that which fills up space" would have no meaning.* On the basis of the general theory of relativity, on the other hand, space as opposed to "what fills space," which is dependent on the coordinates, has no separate existence. Thus a pure gravitational field might have been described in terms of the g_{ik} (as functions of the coordinates), by solution of the gravitational equations. If we imagine the gravitational field, i.e., the functions g_{ik}, to be removed, there does not remain a space of the type (1), but absolutely *nothing*, and also no "topological space." For the functions g_{ik} describe not only the field, but at the same time also the topological and metrical structural properties of the manifold. A space of the type (1), judged from the standpoint of the general theory of relativity, is not a space without field, but a special case of the g_{ik} field, for which—for the coördinate system used, which in itself has no objective significance—the functions g_{ik} have values that do not depend on the coordinates. There is no such thing as an empty space, i.e., a space without field. Space-time does not claim existence on its own, but only as a structural quality of the field.

Thus Descartes was not so far from the truth when he be-

---

* If we consider that which fills space (e.g., the field) to be removed, there still remains the metric space in accordance with (1), which would also determine the inertial behavior of a test body introduced into it.

lieved he must exclude the existence of an empty space. The notion indeed appears absurd, as long as physical reality is seen exclusively in ponderable bodies. It requires the idea of the field as the representative of reality, in combination with the general principle of relativity, to show the true kernel of Descartes' idea; there exists no space "empty of field."

Generalized Theory of Gravitation

The theory of the pure gravitational field on the basis of the general theory of relativity is therefore readily obtainable, because we may be confident that the "field-free" Minkowski-space with its metric in conformity with (1) must satisfy the general laws of field. From this special case the law of gravitation follows by a generalization which is practically free from arbitrariness. The further development of the theory is not so unequivocally determined by the general principle of relativity; it has been attempted in various directions during the last few decades. It is common to all these attempts, to conceive physical reality as a field, and moreover, one which is a generalization of the gravitational field, and in which the field law is a generalization of the law for the pure gravitational field. After long probing I believe that I have now found * the most natural form for this generalization, but I have not yet been able to find out whether this generalized law can stand up against the facts of experience.

The question of the particular field law is secondary in the preceding general considerations. At the present time, the main question is whether a field theory of the kind here contemplated can lead to the goal at all. By this is meant a theory which describes exhaustively physical reality, including four-dimensional space, by a field. The present-day generation of physicists is inclined to answer this question in the negative. In

---

* The generalization can be characterized in the following way. In accordance with its derivation from empty "Minkowski space," the pure gravitational field of the functions g_{ik} has the property of symmetry given by $g_{ik} = g_{ki}$ ($g_{12} = g_{21}$, etc.). The generalized field is of the same kind, but without this property of symmetry. The derivation of the field law is completely analogous to that of the special case of pure gravitation.

conformity with the present form of the quantum theory, it believes that the state of a system cannot be specified directly, but only in an indirect way by a statement of the statistics of the results of measurements attainable on the system. The conviction prevails that the experimentally assured duality (corpuscular and wave structure) can be realized only by such a weakening of the concept of reality. I think that such a far-reaching theoretical renunciation is not for the present justified by our actual knowledge, and that one should not desist from pursuing to the end the path of the relativistic field theory.

Acknowledgment is made to the following for their kind permission to use copyrighted articles in this collection:

Philosophical Library for "On Receiving the One World Award," "The War Is Won, but the Peace Is Not," "Mahatma Gandhi," copyright, 1950, in *Out of My Later Years;*

Scientific American for "On the Generalized Theory of Gravitation," copyright, April, 1950, in the *Scientific American;*

The Library of Living Philosophers, Inc., for "Remarks on Bertrand Russell's Theory of Knowledge," translated by Paul A. Schilpp, copyright, 1951, published by Tudor Publishing Company;

UNESCO for "Culture Must Be One of the Foundations for World Understanding," in the December, 1951, issue of the *Unesco Courier;*

University of Chicago Press for "Dr. Einstein's Mistaken Notions" and "A Reply to the Soviet Scientists," from the February, 1948, issue of the *Bulletin of Atomic Scientists;* "Symptoms of Cultural Decay," from the October, 1952, issue of the *Bulletin of Atomic Scientists;*

British Association for the Advancement of Science for "The Common Language of Science," from Vol. 2, No. 5, *Advancement of Science;*

Monthly Review for "Why Socialism?," from Vol. 1, No. 1, *Monthly Review;*

The Franklin Institute for "Physics and Reality," reprinted from *Journal of the Franklin Institute,* Vol. 221, No. 3, March, 1936, with permission of the Franklin Institute of the State of Pennsylvania;

The American Scholar for "The Military Mentality," reprinted from *The American Scholar,* Summer, 1947;

The Christian Register for "Religion and Science: Irreconcilable?," from the June, 1948, issue of *The Christian Register;*

Harcourt, Brace and Company, Inc., for "On Freedom," from *Freedom: Its Meaning,* edited by Ruth Nanda Anshen, copyright, 1940, by Harcourt, Brace and Company, Inc.;

The American Association for the Advancement of Science for "The State and the Individual Conscience," from *Science,* Vol. 112, p. 760, December 22, 1950; "The Fundaments of Theoretical Physics," from *Science,* Vol. 91, p. 487, May 24, 1940;

The Atlantic Monthly for "Atomic War or Peace," from the November, 1945 and November, 1947, issues of *Atlantic Monthly,* copyright under the titles "Atomic War or Peace, 1945"; "Einstein on Atomic Bomb, 1947," by The Atlantic Monthly Company, Boston, Mass.;

Princeton University Press for "Testimonial Written for *An Essay on the Psychology of Invention in the Mathematical Field,* by Jacques S. Hadamard," copyright by Princeton University Press, 1945;

Collier's for "Why Do They Hate the Jews?," copyright by the Crowell-Collier Publishing Company, *Collier's,* November 26, 1938.

Methuen and Company, Ltd., for "Relativity" from the 1954 revised edition of *Relativity, the Special and General Theory,* translated by Robert W. Lawson; "Geometry and Experience," from *Sidelights of Relativity.*